The Fellowship of Fire

Dan Pence

The Fellowship of Fire

Outskirts Press, Inc.
Denver, Colorado

Outskirts Press, Inc.
http://www.outskirtspress.com

ISBN: 978-1-4327-6793-8

PRINTED IN THE UNITED STATES OF AMERICA

Contents

Preface

Many people choose relative routine, safe and predictable careers. Others end up spending a lifetime at such jobs due to circumstances, family and other constraints. Some actually enjoy it. Yet there are others-people who thrive on the unpredictable; men and women who live for a life on the edge. Some join the military, become law enforcement officers, pilots, structural firefighters, or choose similar occupations.

And some fight wildfires. These adrenalin junkies choose life on the edge, a world of uncertainty and the fellowship that accompanies each fire. This book is dedicated to that fellowship, and especially to the old "ground pounders"; the firefighters who predated the use of helicopters—men who shouldered 80+ pound packs, gathered their tools and hiked to wildfires beyond the reaches of civilization.

CHAPTER 1

Don't Play with Fire

As far as I can recall, my first major incident involving fire occurred in the fall of 1945 when I was six years old. Harold Wick, the local forest ranger's son, was my best friend. Harold and I were obvious victims of a plot hatched by a pile of dry leaves and an available match one fine fall day. It really wasn't our fault. My grandfather had raked dry leaves from an irrigation ditch that ran in front of our house and planned to burn them when he got time. Harold and I found a match, and figured we could save grandpa a lot of trouble.

The resulting fire was quite exciting and we were spreading it to other leaf piles when Mom caught us. Mom wouldn't accept our claims that Spot, the local Springer Spaniel dog, was responsible for the blaze. She skipped Harold home with an impressive switch of her own choosing, where he faced an uncertain fate at the hands of his mother.

For me, the aftermath was not pleasant. This session did include an important message: "If you play with fire, sooner or later you will get burned." Mom was right, as moms usually are, and I have been burned more than a few times. Yet the fascination remains-wildfire is exciting to be around. And there is a special fellowship that goes with the firefighters who challenge wildfires.

My friend Harold's father, U.S. Forest Ranger Johnny Wick, did more than help shape that survival and my future. The local Boy Scout troop operated under the very strict control of the Mormon Church. There were strings attached for gentiles. Assisted by our father and other parents, Ranger Wick formed a Forestry 4-H Club that influenced many of our lives. It resulted in careers for several of us, especially my brothers and me.

No one had heard of poverty in the Big Lost River Valley in those days. It was just a natural way of life. No one had much money or material possessions but that was okay. All we had to do was look around and it was obvious there were others who had less than we had.

Thankfully there were no child labor laws in the 1940's and 1950's. My parents couldn't afford to send five boys to college, but they met our basic needs and taught us to work. We helped in their general store from the time we could write prices on merchandise until we left for college. We worked in the family slaughter house and learned to be butchers. Our parents couldn't afford to pay us wages. We started picking potatoes when we turned eight, then advanced to farm laborers/cowboys the summer we turned thirteen. All of our wages went into our college funds. Life was good and we had no complaints.

Then I was through high school. The U.S. Forest Service

was hiring "smoke chasers" (firefighters who hiked to forest fires) for the summer. Fighting wildfires sounded exciting and I was anxious to try my ability there. And yes, the fires met my expectations and more; as did the bond that develops between the people who step out from normal society to challenge them in the wild lands of our country.

CHAPTER 2

"4-11, I Have a Fire"

I signed on as the Marsh Creek Patrol on the Stanley Ranger District of the Challis National Forest in 1957, the summer following high school graduation and my eighteenth birthday. This was a horseback summer smoke chaser and recreation patrol on the headwaters of the Middle Fork of Idaho's Salmon River. I rode from my tent camp home at Lolo Creek Campground at the end of the road, down Marsh Creek to where it joined Bear Valley Creek to form Idaho's Middle Fork of the Salmon River in what is now the River of No Return Wilderness.

My assignment was to give fire prevention messages to any visitors I encountered, post fire prevention and recreation signs, clean up garbage left by those thoughtless people who were too lazy to pack it out, and to fight any fires encountered along the way.

I also enrolled in the College of Forestry at the University of Idaho for the fall semester, assuming that the money

I saved as a farm laborer/cowboy, supplemented by the summer's Forest Service wages and whatever I earned at any work I could find between classes would pay the bills.

Fire School at the old Civilian Conservation Corps (CCC) Camp on Morgan Creek started my voyage into the world of wildland fire. We learned the fundamentals of fighting wildfires, map reading, locating specific points on the ground by a process called triangulation, and even fought a training fire that was set specifically for us to put out. We learned a lot more on the actual fires that followed.

A major point emphasized throughout the school involved memorization of the "Ten Standard Fire Fighting Orders," and ten "Situations That Shout Watch Out". The latter has been expanded to eighteen now to cover additional situations that have seen firefighters die. We also had to learn the chain of command involved in fighting fires.

Above all, a firefighter is to fight fire aggressively, but safely. There aren't enough acres out there anywhere that is worth the life of one firefighter.

A group of smokejumpers and a smoke chaser had a major disagreement with a fire on Montana's Mann Gulch just eight years earlier, in 1949. Only three out of twenty-eight men involved survived that incident. The standard fire fighting orders and first ten situations that shout watch out resulted from the investigation into why those men died.

In 1957, the Forest Service used a combination of both the "10" and "4" codes for transmitting radio messages. Radio communications were marginal at best in that remote, rugged mountain country. Someone, including the Army in World War II, thought numeric codes were easier to understand than normal speech. Considering the faint radio signals on those antiquated SPF radios standard for Forest Service use, often interrupted by static and other transmissions, they were right. I never found out for sure what the "SPF" acronym actually meant. I assume it referenced the special radio frequency designated for Forest Service use. Unfortunately, atmospheric conditions seriously impacted those low frequency channels and there was a lot of overlap with other broadcasting units and static.

We manned a lot of fire lookouts. Central Idaho is so rugged the lookouts weren't on towers. They were small wooden shacks perched on rock piles on top of jagged peaks. Aerial patrols were expensive and could only spot fires when they were in the air. A lot of that part of Idaho is in designated wilderness now. There weren't a lot of people around to spot fires when they were small. Even with all of the lookouts in place the rugged terrain created large areas that are "blind" as far as the lookouts are concerned. Aerial patrols were commonly launched immediately following lightning storms or when several fires were going.

So we put people in little shacks on mountain tops, maintained ground line telephones to some, and relied on the marginal radio systems to relay messages from others. The telephone lines consisted of a single heavy gage wire generally strung from insulators that were nailed on trees. Those phones cracked and popped, especially during lightning storms. A person could generally shout and get a message through if too many trees hadn't fallen across the line to short it out.

When a lookout spotted a fire, he radioed "4-11 (emergency traffic), I have a fire." Everyone stopped trying to transmit over the radio when they heard "4-11," then strained to hear the lookout's best guess as to the fire's location. He used a compass reading or azimuth to give a bearing from his lookout to the fire's smoke column.

Other lookouts scanned the horizon for some sign of the fire being reported, to help "triangulate" the fire's location. If another lookout saw the smoke and reported his azimuth, it could be pinpointed based on where the two readings crossed.

I was maintaining the telephone line between the Cape Horn and Seafoam Guard Stations with Jack O'Conner, the Cape Horn Guard (the smoke chaser stationed at Cape Horn), one afternoon about a week after fire school when the first call came. We had no radio so the station guard from Stanley simply followed the telephone line until he found us.

Talk about a rush of adrenaline! I cannot put that euphoria into words, other than that we were sure we could whip anything a wildfire had to offer and now we had the opportunity to prove it. We dashed back to Stanley where Ranger Bob Newcomer was gathering everyone he could contact.

As soon as enough people arrived Bob assigned Sam Warren, his assistant, as crew boss and we were off. We rode on benches in the open bed of pickup trucks along with our tools and gear.

It was the first fire for most of us. We were mostly kids, between the ages of eighteen and twenty on our way to declare war on a natural event of nature. I'll never forget the feeling of fellowship and anticipation, and this was just the beginning.

The fire was named Carmen Basin after a landmark on the East Fork of the Salmon River on the adjoining Clayton

Ranger District. God obviously had a lot of material left over when he created the earth. He deposited a lot of it in Idaho's Salmon River country, generally referred to as the Salmon River Breaks. Fires burn well on the almost vertical canyon walls involved.

As fires go, Carmen Basin wasn't that much. Fire season was young with a lot of green grass and winter's residual moisture. We had to cross the East Fork of the Salmon River in a hand propelled cable car to reach the base of the fire. The only smoke chaser injuries I remember from that fire resulted from men getting their fingers tangled up in the pulleys that connected the car to the cable that was strung across the river.

The fire burned savagely through a couple of hundred acres in the previous year's cured grass and a lot of dead and dying timber accompanied by strong winds before it reached the ridge top and the wind died down. Discounting the steep climb through rugged topography, it wasn't that hard to catch. Yet it was the first of many, and an excellent training exercise.

Carmen Basin identified a hazard no one, including Mom, had warned me about. Trees get old, die and fall over. Some of these logs lodge horizontally across steep slopes. Frost, rain, animals and other factors dislodge boulders on rocky hill sides. The rocks roll downhill and frequently lodge behind logs. Fires burn logs. The rocks are suddenly free again. For fire fighters on steep slopes, looking up to see what caused a loud crash above them only to have huge boulders come tearing by at head height is a normal, if somewhat terrifying, part of the job. Related moments are especially interesting at night when a person can't see the rocks.

Yes, Mom, you can get burned playing with fire, but there are other hazards out there that can get you as well.

We were tied in with a group of "pick up" firefighters, assigned to build line around the right flank of the fire. There weren't many well-trained fire crews on most forests then. The Forest Service couldn't afford many employees on their very limited budget. When a large fire started, the ranger simply started hiring people from local bars and other places frequented by unemployed men. These were the "pick ups" I worked with on this and many of the fires that followed.

I knew a lot of the pick ups on Carmen Basin. I had gone to high school with several of them. They were just kids fresh out of high school looking for work until they could decide what to do with the rest of their lives. Other pick ups were town drunks and other older marginally employable men who hung around bars and other places. We built a lot of fire line in spite of a lack of experience and, in some cases, severe hangovers.

Smokejumpers from McCall, Idaho arrived over the fire in a tri-motor Ford airplane as soon as the wind died down that evening. What an impressive sight! Actually, one of them got caught in a wind gust just after he hit the ground and his open parachute dragged him through a bunch of rocks. A couple of us jumped on the open parachute to collapse it.

I learned to love smokejumpers. It was their food. Smoke chasers carried "fire packs", around 80 pound knapsacks, that included three day's worth of World War II combat rations (C-rats) and, if you were lucky, a small chicken feather "mummy" sleeping bag. A cold can of greasy 1944 vintage beef stew and two ultra-dry crackers isn't gourmet dining. Sleeping in the chicken feather mummy bags was a lot like sleeping in a deep freeze in that high country. Both beat the alternative of going hungry with no sleeping bag.

Smokejumpers came with aerial supply drops that included such luxuries as canned hams, fresh fruit, juice, sealed cans of coffee, and other delicacies foreign to us ground pounders who existed on what we carried on our backs.

Besides, when the fire was controlled, smokejumpers left. They were deemed too important in case other fires broke out to stay on a fire once it was controlled. The rest of us had to stay on the line until the fire was mopped up (24 hours after we saw the last smoke). The smokejumpers took their basic tools with them. They abandoned what was left of that delicious, expensive food. Those left to mop up the fire were suppose to bury it. Most of us were college kids with little money. We bought our own groceries at our assigned workplaces between fires. I remember splitting out the thigh seams in my jeans when I stood up with well over a hundred pounds of jumper groceries on my Trapper Nelson pack board on more than one fire.

I gave serious thought to trying out for the jumpers but the rangers I worked for and other permanent Forest Service personnel talked me out of it. They maintained they could give me better training in Forest Service operations if my primary duties weren't just fire. Their discussion made a lot of sense and I still got on a lot of fires.

The rangers lacked funds necessary to complete the many jobs that needed to be done. Fire funds were a bit more versatile and available. We were hired to fight fires when they occurred and were paid out of special fire funds when weather conditions exceeded a specific fire danger level. We did everything from maintaining campgrounds to clearing trails between fires. Since I was a forestry student, they also had me assist in technical range and wildlife studies and in preparing and administering timber sales.

Unfortunately, a close encounter between my left foot and a sharp ax limited my fire experience in 1957. I ended up short one toe with severed bones in two others when my ax bounced off a limb while cleaning a trail. It was a long walk out on a bloody foot.We were over five miles from the road at the time.

My older brother, Ned, due to get out of the University's Forestry Summer Camp shortly after my accident assumed the Marsh Creek Patrol job. I limped around with a walking cast on my left foot, cleaned campgrounds in the morning and manned the radio at the ranger station in the afternoon. How I hated dispatching crews to fires when I couldn't go myself.

The primary social function in Stanley, Idaho during the summer was the "Stanley Stomp". It was booked as a dance at the Jack of Diamonds Club, but routinely degenerated into more than one drunken brawl before daylight. We enjoyed the action, including beer without ID checks and an opportunity to pursue cute co-ed's who worked the area's dude ranches each summer.

I was pretty bashful, the result of having only brothers for reference plus a mother who assured us that girls only wanted to get you off in some dark place where they could do terrible things to your body. I found out later that Mom was right about what the girls wanted to do, although she was wrong about the terrible part. I made a few futile efforts with the ladies. Knowing how to dance and not having a cast on one foot would have helped. Mostly I drank a lot of beer.

Snow in the high country ended fire season by early September. And then there was college.

CHAPTER 3

Clayton Station Guard

Job offers for the next summer included a promotion to station guard for the Clayton Ranger District, thirty some miles down river from Stanley. Station guard is the lead smoke chaser for a district, so it seemed pretty important to me. Brother Ned had the station guard job at Stanley, so I took Clayton. Marv Larson was ranger, backed by his wife, Fern who worked as a part time "Ranger Clerk" and covered things while Marv was gone.Marv was an old "Pony Ranger", which meant that he had worked up through the ranks. He didn't have a college degree in Forestry. Several rangers at that time had attained their rank during World War II for the simple reason that there weren't enough college graduates to fill the need. A lot of them knew more about practical land management needs than some of the college graduates I worked with later. Marv was very good at what he did. He and Fern constituted the entire permanent district staff; the rest of us worked seasonally.

Marv had me clear trails, mark timber, monitor livestock grazing allotments, clean campgrounds, construct and maintain range fences and other improvements, survey homestead corners, install wildlife improvements, work recreation and do anything else that came up between fires. Somewhere along the line Marv had volunteered to have his fire crews build and install outhouses (outdoor toilets) for all of the ranger districts on the Forest. We became experts at constructing and installing outdoor privies.

And there were fires: the woods really burned in 1958 and we were there. The basic pay wasn't much per hour, but we more than paid for college tuition that summer. In those days, the U.S. Forest Service thrived on policies that would violate numerous union, Equal Employment Opportunity and civil rights laws today.

We were put "on call" when the ranger thought there might be a fire on one of our scheduled days off. We received no pay on call. We had to remain where they could contact us, and be en route to a fire within an hour if called.

Next step up was "standby." Standby paid twenty-five percent of our normal rate. Our getaway time for heading for a fire was to be less than fifteen minutes. During normal working hours, our get away time was to be less than five minutes from the time we were notified of the fire dispatch.

There were lots of fires and lightning storms that summer. Since we were almost always either working, on standby, or on call, we didn't stray far. Usually we worked during the day, even when we were on call. Since we couldn't go anywhere, the time passed faster when we worked. I don't remember a day off from July 1 until we returned to college in mid-September.

I loved it. Most people work predictable, routine jobs. We never knew what was going to happen next. Any job could be interrupted at any time by a fire call. I think most fire fighters thrive on this uncertainty, the related fellowship and the adrenalin rush that comes with living on that edge. Marv even burst into the tent where we slept with a fire assignment in the middle of the night on more than one occasion.

There were some false alarms, especially early in the season. Our first came just after we put a new man on Custer Lookout. A pretty hot lightning storm came through but there was also quite a bit of rain with the storm. The lookout called in a fire on the Yankee Fork of the Salmon River.

Keith Slane, Bonanza Guard, led us through some very rugged country looking for it. We had a radio with us and could talk directly to the lookout just across the canyon. We started a test smoke to help him guide us in before dark. He saw our smoke and said we were close.

Torrential rain arrived shortly after dark. We were soaked by midnight, but still no fire. Keith really tied the lookout down and he finally admitted that he had never seen a smoke or fire, just a lightning strike that "--was so big it just had to have started a fire."

We were all soaked and cold by then. We started our own fire. All of the old Army surplus "mummy" sleeping bags had been sent in for cleaning so I had nothing to sleep in. Keith had one he hadn't turned in, along with a small piece of canvas. Keith took the damp bag while I curled up in the canvas. We kept the smoky fire going and tried to ignore the skiff of snow that settled over us before morning. I tried to hang on to a mummy bag after that, even if it was dirty.

One Saturday night we attended the "Stanley Stomp" and had stumbled into our sleeping bags in the mouse infested tent behind the ranger station after two a.m. when the fire call came. The mill at the Livingston Mine, almost two hours away on perilous dirt roads, was on fire.

The Forest Service constituted the total fire control option for rural residents in that part of the world in 1958.

Predictably, the structures were pretty well gone by the time we arrived, but a high wind was driving the fire as a crown fire through adjacent timber. When water was available, we used small two cycle portable pumps manufactured by Pacific Marine Company. They produced an amazing amount of high pressure water. Standing in a cold mountain stream trying to get a Pacific Marine Pump started at four in the morning close to 10,000 feet in elevation is a tough way to work off an evening of beer drinking. We finally discovered that I had forgotten to open the vent on the gas tank and got the pump running right. The buildings were lost by the time we arrived but we saved a lot of trees, thanks to that water.

About this time, I discovered an interesting young lady who worked at Robinson Bar Guest Ranch. My limited contact indicated that she was more than willing to show me essentially everything my mother had warned me about. I was very interested.

We had what promised to be an exciting and educational date lined up for a Stanley Stomp but the Basin Butte fire call came just two hours before that event. I interrupted my five minute getaway time to ask Fern to call Robinson Bar and advise the young lady that I wouldn't make the date.

I never found out whether Fern just had too many things

going, what with the fires and all, or if she simply decided my social life was getting out of hand. I think she was in cahoots with my mother and the girl's aunt who ran the lodge. Anyway, the call was never made and the young lady was left all dressed up with no place to go. I tried to call and apologize when I got back, but she wouldn't answer the phone. It was back to drinking beer for the rest of the summer.

We had to hike about ten miles to reach the base of the Basin Butte Fire and it was dark when we hit the line. The fire was crowning (burning through the tree tops), and moving fast ahead of a brisk wind through dense lodgepole pine on steep terrain. Dead logs littered the ground, and a lot of dead trees (snags) remained standing.

We found ourselves busy cutting fire line through the dense timber and logs in the darkness, heat and smoke. Visibility was nil. We would hear a sudden rush of air followed by a loud crash, and a large log would be lying in a cloud of dust right next to us that hadn't been there an instant before. The roots just burned away and the snags made no sound except for the rush of air as they fell. A lot of them fell that night. Any one of them would have killed a person if it struck them. We built miles of nervous fire line and somehow survived with no casualties. I never told Mom. She still thought you just get burned.

I saw my first "Mae West" when the jumpers came in at daylight. A shroud line that holds the canopy to the jumper looped over one man's parachute as it opened. It did form an interesting resemblance to a very full bra. I also learned a few new swear words from the frantic jumper who kept yanking the lines as he came in. The atmosphere is pretty thin at that elevation and he hit the rocky ground hard.

I don't recall all of his injuries, just that he broke bones and we had to delay evacuation until a bulldozer that was building a road to the fire arrived after noon.

A couple of pickup fire crews arrived on the new road giving us the manpower to make good headway on the fire line. I had gone to high school with two of the pickups and hadn't seen them since we graduated. The three of us slipped into the supply area one evening when a thunderstorm threatened rain and "borrowed" a cargo parachute that could serve as a makeshift tent. I enjoyed catching up on what they had accomplished since high school that evening. We were young and full of life and our current experiences on the fire line formed a special bond.

Both of them had teased me about having too much interest in the outdoors and not enough interest in girls and parties while we were in high school. Now the simple burden of having to make important life decisions that would shape the future seemed to scare them. Yet the fellowship of fire formed a common bond for a great discussion and an enjoyable evening.

I knew that I wanted to be a forest ranger and what I needed to get there. My friends weren't so sure. One wanted to be an artist and the other was still working on the whole concept although he enjoyed working in the outdoors. He was mentally capable of doing anything he wanted. He eventually picked up a special masters program funded by the Forest Service to become a fisheries biologist. He was caught cavorting with a young lady who definitely wasn't a "proper" lady in a guard station one afternoon when he was supposed to be collecting fisheries data in the White Cloud Mountains. God may forgive but the U.S. Forest Service doesn't. I still wonder if I could have said anything that evening that might have helped either of them more with life, although we all have to make our own decisions and live with the consequences.

We even got to haul the groceries the jumpers left off the Basin Butte Fire in the back of Dad's '49 Ford pickup truck brother Ned had rented to the Forest Service that summer.

There were lots of two-man smoke chaser fires between the major blazes in 1958. These were generally small fires (less than an acre). I'd just grab a buddy, my fire pack, and tools, and head for the fire. The country was so undeveloped that we didn't drive to a fire that year. The fire report would come in and we'd be dispatched to drive as close as we could get, put on our fire packs, pick up out tools and head for the fire location on foot. Helicopters hadn't been included in the budget for the Challis National Forest yet.

If we were lucky, we also packed a hand held tube type Motorola radio that added another ten pounds to our gear. If we were even luckier, we could actually talk to a ranger station or lookout with the radio once we got to the fire. We could generally talk to a patrol plane if it was overhead.

We didn't have a lot of radios. We communicated with the patrol plane by ground signals on most smoke chaser fires. We carried rolls of orange crepe paper in our fire pack and laid it out in pre-arranged code. I recall a lot of fires where we spelled out the "SS" code with the orange crepe paper that indicated we needed a chain saw. Those in the plane would see the signal as they flew over us and drop an old hand powered cross-cut saw instead of the requested power saw. The cross cut saw beat worrying our way through large burning trees with a

Pulaski (a cross between an ax and a grubbing hoe). We cussed the people out in the plane, figuring they were too cheap to drop us what we needed. In reality, I suspect gas fumes from a chain saw not only represented a very real safety hazard to those in the plane, and the fumes would add to air sick potential for the occupants since the air can get pretty rough over the mountains in the weather they flew through.

I don't remember a fat smoke chaser or smoker who lasted long. I wish I were close to that physical condition again. We could walk straight up steep mountains carrying eighty pound packs without stopping and rest as we jogged down the other side. Back then, most trail signs in the Middle Fork of the Salmon country were in pack string hours rather than miles. Trail crew foremen rode a horse and led up to eight pack mules with their gear as they worked trails in the area. They clocked how long it took to ride between points while leading the pack string. The time it took is the time that went on the signs, which made a lot of sense since time between points means a lot more than miles in that near-vertical country.

We could throw on our fire packs (weight depended on how much personal gear and water you chose to carry), grab a shovel and Pulaski, and beat the time on those signs by twenty-five percent consistently. We'd take turns carrying the radio, if we had one. Good boots were critical. The fire line ten miles from a road is a poor place for blisters.

Most of us wore White brand jumper boots, hand made in Spokane, Washington. I have tried a lot of boots, but Whites are the only ones that consistently stand up to the punishment that rugged country and the heat from the fires gave them. We'd build up layers of calluses on our feet, but I don't remember wearing a blister in White boots.

The Occupational Safety and Health Agency (OSHA) would have a fit if they knew how much carbon dioxide we ran through our lungs from the smoke we inhaled along the fire line that summer. We didn't realize that was a problem.

A. Bryan Kelso was the only exception to smoking that I met. Bryan was about sixty years old, chain smoked hand rolled Bull Durham cigarette, and climbed mountains like a mountain goat. He was district trail crew foreman and animal packer. I kept Bryan's required diary for him (I don't think he could write), and he taught me to wrangle and pack mules.

And then it was September, with college due to start in a few weeks. It looked as though fire season was over until the Gardner Creek Fire took off. We were the first ones dispatched. It was one killer of a hike into the rugged country on the north end of the White Cloud Mountains. The fire was kicking up quite a column of smoke, and we didn't have a radio. Jim Hintze, a classmate from Mackay High School, was on Lookout Mountain Lookout. Jim looked right down on the fire from his lofty perch. The powers that be didn't like what he was telling them and ordered a helicopter plus smokejumpers from Boise.

The jumper plane went in first, followed by the first helicopter flight carrying smoke chasers from other districts when we were about half way there. We couldn't do anything but keep walking. Supplies necessary to support a hundred-man fire camp were dropped, including some Pacific Marine Pumps and lots of hose since we had access to the water in Gardner Creek.

We completed the long hike to the fire line well after dark to find close to a hundred men who had either been ferried to the fire via helicopter or had jumped. Big brother Ned was there (his first helicopter ride).

It took several days to control the fire, and since it was on the Clayton District we were naturally left behind to mop up while the others were 'coptered from the fire. Ned somehow got left with us, which helped.

Mop up is a firefighter's least preferred job. He has a lot of adrenaline going while the fire is hot and moving. It's another matter when the fire is "controlled". Someone has to patrol the line, making sure nothing remains on fire close to the line that might allow the fire to take off again; tedious, dirty, monotonous work. Yet such a situation helps form a special bond among firefighters, one that can last a lifetime.

Sagebrush/grass fires are easy to mop up, since only small diameter (fine) fuels are present. Everything is consumed by the initial fire. A sage/grass fire may continue to smolder for a day or so in ant hills and animal manure, but that's easy to put out. Timber fires are another matter, especially along canyon bottoms where there's a lot of decaying vegetative material (duff), dead roots, old stumps, and logs to hold the fire. Gardner Creek included a lot of heavy timber, with all of the above. It looked like we'd be there forever.

The weather on Gardner Creek turned unseasonably warm and dry. All but some very basic supplies were flown out on the helicopter. All that was left was what we could bury or pack out on our backs. We were left without a tent or other frills. The last we saw of civilization consisted of a food drop from an old single engine Fairchild airplane.

One cargo chute hung up in the highest spruce tree around. Someone climbed the tree high enough to read a card that said the cargo box contained breakfast. We cut the tree down after missing the point that breakfast was primarily a case of uncooked eggs (instantly scrambled). Fortunately, the fishing for small cutthroat trout was excellent on grass-hoppers and safety pins in Gardner Creek, and "fool hens" (spruce grouse) were plentiful if not in season. We could kill the grouse with a shovel. We ate all right.

Nature helped us mop up. Camp was located on a rocky knoll near timber line above the fire. We had finished supper and were gathered around a camp fire just before dark one evening when the "smokes" were getting few and far between. We had seen a few lightning flashes to the west earlier in the day. A massive black cloud full of wind, lightning, rain and hail suddenly swirled over the ridge and engulfed us. And it seemed determined to dump it all directly on us.

We were instantly soaked to the skin. The air reeked of ozone from the massive barrage of lightning strikes that surrounded us. Thunder was continuous and deafening. LeRoy Kline, a University of Utah Forestry student, was in charge. I'll never forget how pale and wet he looked as he stuttered "D-do-don't y-yo-you t-th-think w-we s-should split up, s-so one s-strike d-do-doesn't get us all?"

We were all scared stiff, but there was no place to go and no time to get there. The storm only lasted for an exciting twenty minutes or so. During that short period it dumped well over an inch of rain and hail and what seemed like a million lightning bolts in our immediate vicinity. Everything, especially our sleeping bags, was drenched. A huge spruce snag fell between Ned's and my sleeping bags. I'm glad we weren't in them at the time. Night time temperatures in September 9500 feet up in the Idaho Rockies aren't real user friendly. We spent most of the night huddled around our smoky, re-kindled campfire, trying to dry things out.

Jim Hintze hiked down from the lookout the next morning to report that a lightning strike started a fire in an old snag a couple of miles away in spite of the moisture. It was reported to be on the Stanley Ranger District side of the line, so Ned took it on alone. It turned out to be just on the Clayton side of the ridge, which didn't bother Ned. He had it wrapped up in short order and headed off the mountain towards Stanley the next day. It took us a day longer to wrap things up on Gardner Creek. At least the hike out was mostly downhill. I just had time to check out and head for my sophomore year at college.

And what an exciting summer it had been. We had met raging infernos in close combat, learned how to deal with them and survive, and had formed a relationship with other firefighters that could not have been forged elsewhere.

CHAPTER 4

Jordan Peak

Resource management majors at the University of Idaho faced a summer camp at McCall, Idaho between their sophomore and junior years. It provided excellent field training in ecology, forest measurements, and engineering, but sure created problems for a poor boy's budget. Fortunately, the camp ended in early July, so we could still get in some work time.

The Forest Service had an agreement with the University stating that the students were available as an emergency fire crew. There were several smoke chaser fires around and we could hear the jumper and retardant planes going out almost every day. I'm sure my buddies and I almost drove the fire dispatchers crazy with our nightly visits to see if they had a fire for us but we didn't get to go.

The camp ended the weekend after the Fourth of July, 1959. We got out at three p.m. that Friday. I headed east over dirt back country roads for Mackay to dump my college gear with my folks, then planned to report at the Clayton Ranger Station the following Monday. I had to drive by the ranger station en route to Mackay. It was about ten p.m., and the lights were on at the fire warehouse. I backed up and drove in to see what was going on. Marv saw me coming and handed me a fire pack. A sheep herder had reported a fire on Jordan Peak. Marv only had a couple of people to send. He put me in charge although I hadn't officially "signed in" yet. Marv assured me that we would handle the paper stuff after I returned. I stuffed a mummy bag, a change of shorts and a couple of relatively clean pairs of socks in the pack, filled the canteens, gathered the others, and headed for Jordan Peak.

We didn't always fill our canteens if we knew there was going to be water along the way since each pint of water weighs a pound. We just drank out of the streams we encountered. I'm sure Guardia and other evil things were present, but we had drank water from open streams all of our lives. Maybe we had grown a tolerance to the evil things that were there.

We met the sheep herder at the first switchback on the road over Loon Creek Summit in the middle of the night. He looked like a tough old boy, sixty or so, lean and mean enough to set out the night hunkering under a Douglas-fir tree high up on a mountain. I stepped into my fire pack, gathered up my tools and radio, lined up the others and we took off up the mountain. More specifically, we spent the next couple of hours trying to stay in sight of one grizzled old sheep herder as he trotted up the steep mountain through the rocks, timber and dark night.

I had spent the previous ten months either at college in Moscow, Idaho, or in the Cellar Bar in McCall, Idaho. That old reprobate had spent most of that time chasing a band of sheep around near timberline in the Idaho Rockies. I was too proud to ask for a breather. Today I can admit that old S.O.B. damn near killed at least one college boy. Thank God he finally got us to a point where we could see the fire, then left to get back to his sheep!

Jordan Peak Fire was the first of an almost continuous sequence that summer. As I recall, I spent forty-two of the forty-five days before school started somewhere on the fire line. Since it was mid-summer before I could report for work, Johnny Pritchett had been appointed station guard. I just filled in wherever I could. One thing probably wasn't "fair" for Johnny. Since he was station guard, his responsibility was the Clayton District if a lot of fires were burning. I had no such constraint, and could head anywhere they needed help.

I had established slightly more than a passing acquaintance with a very pretty physical education major at the university by then. I'm not sure she understood what she did to a young man's constitution whenever she showed up in a sweater and jeans, but she sure had me interested. She spent summers working in the hay fields on ranches near Richfield, Idaho. She promised to show up at one of the Stanley Stomps with some girl friends. That got my fellow firefighters interested as well. The girls made it, we didn't.

CHAPTER 5

Hat Creek and Mystery Creek

I don't recall the specific order of the larger fires that summer. Johnny and I shared bunk space in the back of the office, a distinct improvement over the mouse filled, leaky old tent with outdoor plumbing that served as "home" the previous summer.

A major earthquake hit the upper Madison River near Yellowstone Park in Montana one night. I'd just returned from a fire and was pretty tired. I don't remember waking up although Johnny assured me that we discussed the earthquake. The quake triggered massive earth slides onto campgrounds filled with campers and closed major roads. A lot of people were buried and never found. They were calling for crews to search for survivors next morning. We were waiting to see if we needed to go when a call came in for the Hat Creek Fire on the Salmon National Forest. I was dispatched as crew boss with a crew from the Challis National Forest. We rode as far as we could in the back of a stock truck designed to haul horses and mules.

I was impressed when we went through a settlement that didn't show on maps en route to the fire. We had been traveling marginal dirt roads for what seemed like hours when we came over a ridge and looked down on several shanties on an isolated ranch. Women and kids appeared to be running everywhere when they saw us coming. By the time we got to the settlement there were only some pretty rough looking men leaning in doorways to be seen. We drove on through. They had a lot of old cars around, most of them jacked up and showing outdated Arizona or New Mexico license plates.I found out later that federal agents had made a major raid on a group of polygamous Mormons in the four corners country of Utah, Colorado, Arizona and New Mexico a few years before. Some of those who escaped took over this remote ranch and continued their chosen lifestyle.

We finally reached the end of the wheel track that served as a road where we were met by a helicopter! It was the first time I had ever been airborne. The ride was exciting, but so was the huge convection column of smoke that became increasingly ominous as we approached. We were supposed to stop that thing?

We tied in with other crews and hit the fire line. We were assigned to construct line across the bottom then start up the east flank. We were backed off almost immediately by a terrifying shower of boulders and logs from above. Investigation revealed that a bulldozer had arrived by a different route and was working the steep slope above us without supervision. His line wasn't tied in with anything and wouldn't hold if the fire started moving. A few men were hit by rocks. I don't know why someone wasn't killed.

We got the dozer corralled, regrouped, and took off again. That was one hot fire. We worked right against the fire building direct line until close to midnight when they told us to bed down. We wolfed down combat rations we had in our fire packs and hit the sack near the fire line. We were working on a steep hillside, but we carved out some relatively flat spots with our tool and crawled into our mummy bags (no pad) and got what sleep we could. I had just crawled in when a major earth tremor hit. I remember thinking "Huh, another earthquake," and then it was morning.

Slopes with south and west exposures normally are the driest and burn hottest. We faced a lot of scattered ponderosa pine with dry grass under story on south exposures on Hat Creek. It didn't take long to find out how fast fire can move through that vegetation on steep slopes. All we could do was get out of the way when that stuff took off.

One of the crews found an old military trainer air plane that had crashed through some trees. Two skeletons were still in their seats. It crashed during World War II and had never been found.

Hat Creek was a tough fire to catch, but we managed to do so in about a week. We didn't have to stay for mop up since it was on the Salmon National Forest. Roads were constructed to get equipment to the fire, so we returned to Clayton in trucks.

I had time to restock my fire pack when the Mystery Lake Fire took off on the Challis District. I was assigned as crew boss for our district crew. The biggest problem in getting started involved locating everyone. Our trail crew had just returned from a ten day on shift the night before and had the following four days off. A couple of the younger members celebrated the event by trying to drink Campbell's Bar in Clayton dry the night before the fire call. They were still sleeping the evening's activities off in a tent.

The hike up Mystery Creek obviously was murder on a hangover, but they made it. Fuels on Mystery Creek were just the opposite from Hat Creek with sub-alpine fir, Engelmann spruce and limber pine in a high cirque basin that had been carved out by glacial action just below timber line. Both spruce and sub-alpine fir host very flammable branches right to the ground. When exposed to fire they simply explode or "crown out". More importantly, whenever one of the trees crowns it showers the entire vicinity with sparks and burning debris.

There's little soil at that elevation so it was a tough fire to put out. Timber grew in clumps wherever it could get a foothold in the rocks. The fire simply jumped from clump to clump because of the sparks thrown up whenever a tree crowned. We never knew where the fire was going to show up next.

A lot of the trees were rotten inside. The fire would manage to get into the rotten core and smolder. We had to fall the trees to get at the fire, hopefully before the tree "crowned". One day we found smoke coming from a broken top in a large spruce that hadn't crowned yet. I started to cut it down with a chain saw, hoping we could get the fire put out before it threw sparks into the surrounding trees. Apparently the fire wasn't getting enough oxygen. The tree literally exploded in flames as soon as the saw cut into the rotten core so oxygen could get in. I escaped, minus eye brows and a lot of other hair, and even managed to bring the saw with me.

Normally firefighters throw dirt on a fire to smother it. I doubt that there is a shovel full of dirt on the entire length of Mystery Creek; there was no shortage of rock. We ordered portable pumps that were dropped from the air. We set the pumps up in Mystery Lake and ran the linen fire hose down the slope to fight the fire. Elevation differences were so great that gravity created very efficient siphons. The pumps would idle along, but pressure downhill at the nozzles was so great one man had trouble holding on to a hose. The pressure got so great that it started blowing hoses. We found a smaller lake farther down the hill so we could reduce the pressure.

All of our supplies were dropped from the air. Abbott Flying Service in Challis had an old single engine Fairchild that could really handle that high country. We'd hear "wow-wow-wow" and that old piece of corrugated metal would come cranking over a nearby ridge and start dropping supplies.

More fragile supplies and food came attached to cargo chutes. Rather than set up a kitchen on each fire to cook food on the line, a lot of meals were prepared in kitchens and restaurants in nearby communities, put in insulated five gallon tins, and dropped to fire camps.

Less fragile items, like sleeping bags, were just tied in bundles and dropped free-fall. The cargo chutes didn't always open sufficiently to fill with air so a fair percentage of the supplies on them tended to free fall as well. The parachutes simply trailed along those fast moving missiles looking like a tail. Nothing like a shower from twenty gallons of hot coffee on a streamer chute to get your attention!

It was hard to keep the pick up firefighters out of

the drop zone. Some of them weren't overly bright and thought anything they got to first was theirs. The bundles of sleeping bags tended to explode when they slammed into the ground. I never saw anyone killed on a fire by a sleeping bag or coffee bomb, but it was close on more than one occasion.

Larger fires in these remote locations were served by a "fire camp" that was set up in a relatively safe location near the fire line. People with blisters on their feet or other physical factors that limited their usefulness on the line were normally assigned to "camp crews". Their duties involved such important assignments as digging "slit trench" toilets to serve the camp. The "toilets" were pretty primitive sanitation stations. Those assigned the task simply dug a trench between two trees, nailed or wired a pole between the trees as a seat and hung a roll of toilet paper on a branch or nail. The pole seat was usually an evergreen sapling cut and peeled on site. We all ended up with a pretty sticky streak across the base of our backsides from sitting on the pitchy poles.

Lloyd "Steelhead" Sammons, camp boss on Mystery Creek, ran a darn tight camp. Steelhead had worked his way up through the Civilian Conservation Corps (CCC's). Somewhere along the line he married one of my cousins. He was the first one up every morning. We apparently were first in line on Abbott's drop assignments. The plane would show up right at the crack of dawn. Steelhead had a special knack in hearing the plane coming. He'd start banging on a wash basin and shouting "Awright men, drop your cocks and grab you socks! Rise and shine before breakfast hits you in the head!" Then it would be "wow-wow-wow", the Fairchild would be dropping and we'd be running to get out of the way. How that pilot could work that old ship through the cliffs and peaks in that high country remains a mystery to me. He'd make as many runs as he needed to serve us before heading for the next fire up the line.

We helicoptered off of Mystery Creek. Civilization was catching up to the Challis National Forest but the old Hiller helicopters available to us were not suited for the high elevations on the Challis. A helicopter landing spot (helispot) had been hacked out on top of a cliff. Two firefighters got in next to the pilot, he jumped that ship off the cliff, and we'd start to fall. How far we fell before the pilot gained control depended on temperature, combined weight of occupants, fuel and gear, and how much attention God happened to be paying at the time. We'd finally get up enough speed and the pilot would pull out of that suicide dive and drop the gear and passengers off on Loon Creek Summit for the long drive home. I couldn't get excited about carnival rides after that flight.

We survived the helicopter ride on Mystery Creek, but too many firefighters were killed elsewhere by such maneuvers and the process was revised. A concept called density altitude came into play later, which regulates the type of helicopters we could use and the load that can be carried at different elevations and climatic conditions.

CHAPTER 6

Birch Creek

I'd just gotten my fire pack re-packed back at the station when a car stopped to report that a state highway crew working on Highway 93 at the mouth of East Fork of the Salmon River could see smoke up under Poverty Flats. Everything is relative along the Salmon River. There was quite a bit of poverty but not much flat. A broken ridge line east of Clayton had hosted some marginal mining activity and it was the closest thing to a flat up there-hence Poverty Flats. There wasn't any water in the vicinity, just a lot of old blue beer and whiskey bottles around broken down, abandoned log cabins.

The fire was burning in sagebrush on the steep south face just under the top. It was about an acre in size when another smoke chaser and I finally huffed our way up to it. We dropped our fire packs in a grassy spot just under the fire and hit the line. Things went well in spite of a brisk west wind. We cut line across the uphill (north) side, stopped the lead on the east end and were just starting south and west across the bottom to tie things off when the wind suddenly shifted to the northwest.

And what a wind! We had been warned about cold fronts and the winds that came with them, but no one had predicted this one. A cold front in the northern Rocky Mountains in mid-summer is another relative thing. We had clawed our way up to the fire in ninety degree temperature. The frontal activity to the west accounted for the brisk southwest wind we encountered when we reached the fire. Arrival of the cold front only dropped the temperature a couple of degrees but triggered an immediate and violent ninety-degree wind shift. Wind increases the amount available oxygen and bends the flames towards unburned fuels.

Things went to hell in a hurry. Fortunately, we were building direct line right against the fire and simply had to step through the flames into the "black" (the sagebrush-grass area already burned by the fire). Topography, geologic features, and position on slopes do strange things to the wind. We were just north of the junction of two major rivers (main Salmon and the East Fork of the Salmon) on the upper third of an extremely steep slope surrounded by all sorts of minor side canyons on a bend in the rivers. The fire followed the wind and the wind went crazy.

I've developed a lot of respect for fires in those light fuels. Firefighters can see a fire building up in heavy timber and generally have time to get out of the way unless they've done something stupid. It's another story in lighter fuels such as sagebrush, grass and the chaparral fields common in the southwest. Those fires can appear calm one moment, then turn on firefighters like a snake and eat their lunch in an instant. It's easy to get trapped if you aren't working right against the fire so you can step into the black.

This fire swooped down slope to gobble up our fire packs, then around the canyon face to the south, east and north. Whole slopes simply exploded. Lack of oxygen was the only thing that slowed the fire. Acres of fuel would ignite with a roar and massive cloud of black smoke. It would pause for a few seconds until the wind drove in more oxygen and make the next jump.

The fire was fascinating to watch which was all the two of us could accomplish. Black smoke rolled up the steep draws into the dense smoke cloud or convection column that towered over the entire area. Flames kept shooting high into the convection column, accompanied by a loud roar. Standing next to a building convection column is interesting. Heat from the fire was generating volatile gases from the vegetation as it advanced. Not all of the gases ignited immediately due to the lack of oxygen. These gases would rise until they mixed with sufficient oxygen and would ignite with a roar sometimes a hundred feet in the sky.The steep side canyons and draws on the fire acted like natural chimneys. The convection column swept up them just like smoke in a stove chimney. It would have been suicide to have been in one of those steep draws, or chimneys, ahead of that fire. They exploded as those volatile gases ignited. I had carried a radio up the hill, but, as usual, could talk to no one. The huge mushroom shaped convection column towering over the fire looked like an atomic bomb had gone off. We stood in the ashes where we were safe until things cooled down in the vicinity of our fire packs, then went down to see what we could salvage.

We shouldn't have bothered. Loss of three day's worth of C rations and a chicken feather mummy bag wasn't all that critical, but loss of half a dozen pairs of relatively clean shorts and socks was a catastrophe when that's all you own. Worse yet, I had purchased a small telescoping metal fishing rod and added a reel and some lures just in case I got caught on another Gardner Creek fire situation. They were just a small puddle of melted metal.

Several crews of firefighters responded to the very visible, if somewhat embarrassing, smoke column. I had told the highway crew to pass word on to the ranger station that we should be able to handle the fire when we left the road since I couldn't get out on the radio. Now the fire had expanded to several hundred acres and we were in bad need of help. About a hundred firefighters showed up that night, the wind eventually died down as the cold front passed and we had things wrapped up in a couple of days.

There were all sorts of other smoke chaser fires that summer. We even got to drive to a pretty fair sized one above the Clayton Silver Mine. We received some moisture from the lightning storm that started the fire in a sagebrush opening surrounded by Douglas-fir timber. The temperature had increased and the wind dried the finer sagebrush and grass fuel. The fire burned up the sagebrush and grass opening then went out against the more moist fuel in the timber about the time we arrived.

Then it was back to college to learn all about resource management, and to see if that cute physical education major had really noticed that I hadn't made it to the Stanley Stomp.

CHAPTER 7

Eccles Creek

I returned to Clayton for the summer of 1960 to find an increasing workload. Marv still couldn't afford a full time assistant, but really needed help. He assigned me as "Ranger Alternate," a rather fancy title for a seasonal assistant. A lot of districts had ranger alternates who worked from whenever the snow melted so they could get into the back country in the spring until snow or lack of funding drove them out of the mountains in the fall. Most ranger alternates were technicians with no college training who worked up from various labor positions. They brought with them unique skills in working with horses and mules, running equipment, building structures, and in understanding the loggers, miners and ranchers they had to work with. Most of them had good fire backgrounds. Marv said he was especially happy to have me. He knew that I had a working knowledge of all the above and he didn't need to feel bad when he had to lay a family man off when the budget ran out. He knew he was getting rid of me when school started and no family was involved.

I did about anything and everything there was to do on a ranger district that summer, including a lot of paper work. It was excellent training for a career with the U.S. Forest Service. I did a lot of surveying to locate homestead corners and corners on mining claims that had passed into private ownership under the 1872 mining laws.

Things had been going pretty well at college. I had even gotten brave enough to hold hands with the physical education major. That young lady sure knew the right moves, although they hadn't included any of the interesting things Mom had warned me about.I had essentially stood her up for a movie date one spring Saturday when I had a flat tire en route back from a fishing trip at Coeur d' Alene Lake. I called as soon as I got back to apologize. She responded that she not only understood, but asked how the fishing had been. She even asked if I would take her with me on the next fishing trip! Boy, did she know how to set the hook. We went to the late show.

I had just about worked up enough courage to try to sneak a kiss when the semester ended. Unfortunately, she chose to teach canoeing at a girl's camp out of Coeur d' Alene that summer. New romantic adventures were destined to wait until fall.

There was an old sawmill site on the National Forest on Slate Creek with lots of old slabs, rotten logs, sawdust and tumbled down sheds that were unsightly, and an invitation to unauthorized occupancy by some of the undesirables who frequented the area. The whole mess represented a fire hazard. Marv waited until after fall snow storms arrived and set things on fire a few days before Thanksgiving in

1959. The fire was still smoldering through remnants of the sawdust pile when I reported for work in June, 1960. We had to turn the whole pile over by hand, and eventually constructed irrigation ditches from the creek to soak the sawdust. We saw our last smoke on the Fourth of July.

The Payette National Forest, headquartered out of McCall, Idaho had some serious fires going by then. I was dispatched as an unassigned Crew Boss for the Eccles Creek Fire in Hells Canyon. This meant that I would be assigned a crew after I reached the fire. We flew out of Challis in a small Cessna-my first ride in a fixed wing aircraft.

The 1960 University of Idaho summer camp students were on the fire. I knew most of them and caught a lot of static over my assignment from them. They were a good, if somewhat motley, crew who learned a lot more about fighting fire than about ecology, engineering and forest measurements that summer. They also left summer camp with a much fatter bank account than I had the previous year.

A crew boss is normally assigned about twenty men. I got eighty farm laborers out of a camp in Nyssa, Oregon. They were primarily Hispanic, with a few African-American, migrant farm workers who labored in the sugar beets and other fields along the Snake River. Several of them were from Mexico, working on farm labor permits.

Anastacio Garcia came with the crew as an internal foreman, with several straw bosses under him. I was used to unorganized pick up firefighters who came without any internal organization. Normal pick ups weren't real certain about how hard they wanted to work, which can be a real test for a young college kid crew boss. These men knew how to work and were used to working together. For a pick up crew, they were mighty fine firefighters. Garcia was a lot of help. The need for organized crews was becoming very apparent. Since these farm laborers were already organized for work, they were a natural. All they really needed was some fire training. They became known as SRV (Snake River Valley) crews on the fire line. I learned a lot of Spanish that trip, especially the swear words.

Hells Canyon is well named. It is the deepest canyon in the continental United States, and one of the deepest (and steepest) in the world. Eckles Creek Fire had started along the Snake River at the bottom of Hells Canyon, jumped the river and was burning on both sides of the river in Idaho and Oregon. We worked the Idaho side.

We'd load up on benches in the back of two and a half ton flat bed trucks that came with the farm labor crews each morning, bounce along the questionable dusty wheel track leading to the vicinity of a place called Lockwood Saddle, off load and climb down into the rocky canyon below.

The normal routine was to see how much line had survived the night, pick up on anything that remained or find a new "anchor," then start constructing fire line. Firefighters don't just start constructing line. It has to be anchored to something so the fire can't flank them. In this case, we frequently tied in with line another crew was constructing in the opposite direction from where we were going.

Several of the ten standard firefighting orders came into question here. We were approaching a major fire from above (fires love to run uphill, and can do so much faster than people can run), we really didn't have a good feel for what was happening down below, we lacked good communications, and it was dryer and hotter than hell in Hells Canyon.

These points didn't go unappreciated by the higher overhead. We would just get some good line constructed and things would look good until about two p.m. when the wind would pick up, trees would start to crown, and the fire would start to move. A sector boss (next step above me) would come running down the line shouting "Haul ass boys, she's gonna blow!"

It really wasn't that simple of an order, even after we translated it into Spanish. Hauling ass meant leaving a fire line that held a lot of blood, sweat and tears to crawl back up a near vertical slope in 100+ degree temperatures.

If we were really lucky, we'd reach the road before things went totally to hell. The fire would jump into the tree crowns in various draws, belch lots of smoke and run for the ridge miles above. We wouldn't have quite as far to crawl to reach the fire the next day.

We'd stay at the road just in case an arm of the fire reached it (normally something did), and try to hold. Then it was onto those farm trucks and back to camp near an old mining camp named Cuprum, Idaho.

The trucks were owned by the Hispanics who drove them. Unfortunately, they were paid by the mile. The owners would do anything to save a nickel's worth of gas. The roads were best suited for four-wheel drive vehicles in low gear. Every time the drivers came to a sustained downhill run, and there were lots of them, they'd hit "Mexican overdrive." That is, they would turn the engine off, kick the truck out of gear and coast down the hills. I like to think I'm as brave as the next guy. However, doing thirty miles an hour down a rocky, dusty wheel track surrounded by big trees and cliffs on fifteen to twenty percent grades while trying to hold on to the bench in the back of a two and a half ton farm truck with twenty other men seemed like a fast trip to the Pearly Gates to me.

Several of us worked that safety problem up the chain of command. It became a major emphasis item at every morning briefing. Apparently it lost a lot in translation. If anything, the trip back to camp each evening became increasingly wild. The drivers started racing each other. We finally had to start firing drivers before the point sank in. Fortunately, we didn't kill a crew en route.

The wind really picked up late one evening. The fire made a major run at Lockwood Saddle shortly after my crew headed for camp. The saddle formed the break between divisions established on the fire line. Several crews were still in the vicinity, so they didn't send us back.

Supper that night consisted of greasy pork chops in gravy, cooked in McCall, packed in insulated five gallon cans, and trucked to the fire. The camp crew was reasonably frugal with helpings, since they figured they needed to feed the line crews still trying to hold the fire at Lockwood Saddle. The fire crossed the road, cutting off some of the crews. They went to another camp farther out the road for supper and the night.

By this time, the pork chops had been slowly cooling, hours from anything resembling cooking temperatures. Apparently, some of the SRV folks thought it was a shame to let anything resembling food go to waste. They slipped into the feeding area and made off with the insulated cans of pork chops in the dark.

The high winds that drove the fire through Lockwood Saddle were the forerunner of a major front. Rain hit camp about midnight, coming down in sheets. All we had for cover were a few spruce trees.

By this time the pilfered pork chops had hit bottom. Somewhere in that gravy, grease and time some impressive organisms managed to multiply. The space under each spruce tree was more than fully occupied by several SRV firefighters with their pants down emitting a stream of sharp noises that included an entire education in Spanish swear words that were uttered around retches and other strange noises. I avoided the spruce trees and got a pretty good cold shower.

The rain immediately turned the dust into impassable mud. Fortunately none of the men with food poisoning died. We'd have had to bury them there. There was no way to get them to more medical attention than the first aid we could offer. Those trees should be mighty tall by now since they received a lot of fertilizer from both ends of some sick firefighters that night.

That storm took the fire out of Eckles Creek. Flooding became a problem and several roads washed out. I was demobilized as soon as things dried out enough so we could get out. I flew back to Challis from McCall in one of the old 1930 vintage tri-motor Ford airplanes that were still around. They were very effective airplanes for the mountain flying required by the Forest Service in those days.

CHAPTER 8

Warm Springs

Little brother Lew had turned eighteen the previous November, and graduated from high school that spring. He was assigned as smoke chaser/lookout on Fly Peak on the Challis District. It was a unique job since whoever was stationed there served as both a lookout and as a smoke chaser. The lookout was located on a primitive wheel track that extended from Twin Peaks Lookout towards Sleeping Deer Lookout on the long ridge that separates the Loon Creek and Camas Creek drainages, two major tributaries to the Middle Fork of the Salmon River.

Fly Peak hadn't been manned since World War II. The ranger decided to man it to give additional coverage during bad fire years. Lew had Dad's old 1949 Ford truck that summer. When he, or another lookout, spotted a fire that Lew could reasonably hike to from the road, he took initial attack action alone.

Lew took on a lot of fires that summer. Then he spotted

the Warm Springs Fire on the Challis District, just across from where Warm Springs Creek runs into Loon Creek. It started to burn up a lot of country. It would have been a long hike for Lew, and he had a commanding view of the area from his lookout. He was told to stay on the lookout and keep everyone updated on fire activity. I was dispatched as crew boss with the crew we assembled at Clayton.

We helicoptered from the road near Fly Peak and landed on a meadow near the creek junction. The fire was in some dry fuel covering steep slopes across Loon Creek, and had a lot of potential. A pretty sizeable crew of smoke chasers and smokejumpers were assigned. My crew waded across Loon Creek and cut fire line up the west flank of the fire. It was good line, and it stopped fire spread in that direction.

The major front on the fire had burned into a series of cliffs so we couldn't try to pinch it off by the preferable direct attack. We were directed to continue above the cliffs and build indirect line across above the fire then start a backfire to burn out the fuels between our line and the fire: a plan with some serious flaws. I knew that the fire would have an excellent run at our line if it burned through the cliffs before we got the line completed and had the fuels between our line and the cliffs burned out. The only safety zone I could identify was the ridge where our line was holding on the west side of the fire. It would be a long, dangerous run if we got in trouble. I was young and relatively inexperienced. I chose to believe that those directing us had considered all of the potential problems I could see so it was safe to proceed.

Everything went well until we reached an avalanche chute above the cliffs. The fire was building up pretty well in mountain mahogany and other fuels in the cliffs. I wanted to complete the line as fast as we could, then try to burn it out before things got too serious. We had to construct line through a lot of logs on the ground as we proceeded. I was assigned an unemployed logger with a chain saw to cut out the logs so we could complete line construction. He had nothing to cut when we reached the avalanche chute above the cliffs. I sent him across the chute to start sawing on the other side. That's the only time I ever split someone away from a crew like that.

The avalanche chute bothered me. It was obvious that it could act like a chimney if we got in trouble. Yet our assignment was to cross it. We had just started into the chute when things came apart. The fire crowned a bunch of trees in the cliffs, threw sparks all around us, and came up through the dry grass in the chute like an angry, noisy dragon. The sawyer had stopped in the chute for a smoke. Why anyone would want a smoke, especially there and then, is beyond me. It was a very short smoke break. I didn't need to order the crew to run for the ridge behind us. Last I saw of the sawyer was his backside as he sprinted for the far side of the chute. Then it was fire, smoke and a hard run for safety. We had sparks and burning debris (fire brands) dropping all around us. A major spot fire flared up just below our fire line before we reached the ridge and safety. Flames licked across the line and ignited a stand of Douglas-fir. They burst into a serious crown fire like someone had poured gasoline on them. Fortunately, we were able to jump below the line and run under that blaze before the main fire closed in from below. We reached the ridge none too soon. Everything behind us went up in flames. I counted noses and could account for everyone but the sawyer. There was no way to get around the flames to see if he made it. We couldn't do anything as long as the fire was running up the slope like that.

Lew had been listening to radio messages on his lookout as he watched the fire explode. He was convinced that none of us could have made it out. I took the crew back down to camp and tried to convince myself that the sawyer really wasn't part of the black smoke coming off that mountain. I've never been more concerned about a firefighter and spent some anxious moments trying to analyze what had gone wrong. We simply should not have been in a natural chimney like that on the uphill side of an active fire! There are no resource values that could justify the risk. I was never as happy in my life as I was when one very dirty sawyer came walking out of the smoke on the northeastern side of the fire. He apologized for losing his saw. Saws can be replaced, human lives can't. I almost kissed him. Looking back, that's as close as I ever came to losing someone on the fire line. It would have been my fault for agreeing to go into the chute.

Like a lot of Salmon River Break fires, Warm Springs made a major run that was impossible to stop. Once it reached a major ridge it encountered different winds and settled down so we could catch it.

Mostly firefighters just get dirty on the fire line. Sweat, dust, smoke, charcoal and ashes make an interesting mix. On the other hand, even a firefighter can only get so dirty. After a week or so without a shower, our skin would start to flake off taking the dirt with it. We'd just get used to looking at dirty, blackened skin when shockingly white spots started to show up. We'd just get that spot black, and another big piece of skin would flake off. We looked a bit strange at times, especially if we had been out for more than a couple of weeks.

Warm Springs had one major amenity missing on every fire I had been on up to this point. On most fires we were hard pressed to have enough water to drink, let alone wash our face. We had warm water on Warm Springs!

The warm water flowed from a spring just below fire camp and dropped over a steep bank into Loon Creek. Some years earlier, a mule packer had stuffed a stove pipe into the warm water just as it broke over the bank. What a shower!

That was the good part. The bad part was that we had over a hundred men in camp, and only one shower. My crew was working the farthest from camp and was usually the last to return. We did have a choice. We could stand in the long line for the shower (temperature about a hundred and five degrees F.) or jump into Loon Creek (temperature about forty degrees F. on a warm day).

Folks working closer to camp generally got the shower while we got the creek. I hesitate to guess how many fish turned belly up downstream from camp. When we jumped into that cold stream after a hot, dirty fourteen hour shift on the fire line, we immediately developed a new appreciation for an old story about brass monkeys. There were definite threats to one's manhood, but at least we had a chance to get clean.

We had a lot of smoke chaser fires that summer, but they were getting pretty routine to me by 1960. Then it was fall and time to get back to school for the big senior year. There was also the possibility that I could sneak in a kiss or two from that physical education major.

CHAPTER 9

A New Career and Corn Creek

The U.S. Forest Service was hiring new full time employees in 1961 and I had several job offers. Choosing which one to take now depended on whether Lois Proctor, my physical education major friend, could get a teaching job in the vicinity. She was offered a job teaching girl's physical education and health in Salmon, Idaho. I accepted a job as assistant to the forest timber staff officer on the Salmon National Forest. Lois and I were married April 9, 1961, the first Saturday of our senior year spring break.

We were a couple of pretty excited kids when we arrived in Salmon on June 11, 1961. I had two aunts in Salmon who helped us find a furnished one room shack we could afford to rent. We could haul everything we owned in the back of our well used '56 Plymouth, so there wasn't much to moving.

It was great to have real jobs. Unfortunately, it took a while before the pay checks started showing up. I joke about having married Lois for her money. Lord knows, I didn't have any after four years of college, plus paying the doctor bills when my father had a stroke in 1959. Unfortunately, Lois didn't have any money either. We broke out the checkbooks after the dust settled from the move and counted what we had in the bank. Lois had seventeen cents and I had only a couple of dollars more. Things were pretty tight for that first month or so. Then the pay checks started, and we didn't have to save them to go back to school! Lois' father helped us purchase a new four-wheel drive International Scout and we set out to enjoy the Salmon River Country.

At work I had to memorize the Forest Service timber management manuals and handbooks, existing contracts, and occasionally I even got to help set up some timber sales. My boss, Joe Ladle, had set up some excellent on-the-job training projects to help me understand how the different Forest Service activities fit together. My previous experience at Clayton helped a lot.

They were also impressed by my fire experience. The Salmon includes some vegetation that burns even better than what I had encountered on the Challis. The fires started early on the Salmon in 1961. They had a couple of old converted TBM's (World War II Navy torpedo bombers) that dropped "bentonite" fire retardant on fires. The fire retardant they dropped was a simple mixture of water with some bentonite clay added to help keep the water from evaporating and get it to cling to vegetation like the thin mud the mixture created. A red dye was added so the pilots could see what they hit when they dropped the retardant.

We could hear the planes take off from our house. I'm

sure I almost drove Joe Denny, the fire dispatcher, and his assistant, Bill Wing, crazy by showing up every time we heard a TBM go out.

Timber management was my primary assignment so they couldn't send me out until things really started burning. Once in a while I'd get to run supplies out to a fire somewhere during the night or on weekends. I really wanted a shot at the line.

We had some interesting training that spring. Helicopters were becoming very common on fires in central Idaho by 1961. Unfortunately, the vertical nature of the Salmon River country severely limits landing opportunities. Someone came up with the idea that copters could just hover, and the firefighters could jump out. That seems like a reasonable assumption, but a lot can go wrong when you try it.

We were using small Bell and Hiller helicopters that were not well suited for high elevation flying, especially in hot weather. At best, they could carry two firefighters and their gear. It took a very sharp pilot to make the immediate adjustments necessary for the sudden weight shift on one side of the ship when someone jumped out. The firefighter crawled out of the seat and perched on the landing skid as the aircraft approached the slope where he was to jump. We were supposed to watch the pilot and not the ground. When the pilot was comfortable with his command of the ship, he nodded his head and the firefighter jumped immediately without looking at the ground. I'm basically a trusting person but there are limits. In theory, the pilot started adjusting his flight as he nodded, so the danger was very real that the ship would crash if we didn't jump on cue. One of the first things I noticed was how often pilots tend to nod their heads, whether they are aware of it or not. I did everything I could to watch the pilot and the ground at the same time. We were to tuck and roll as we hit the ground. The tuck was easy enough, but try to stop the roll on a ninety percent slope! I spent about a week digging thorns out of a knee after landing in a prickly pear cactus.

Lois and I were adjusting quite well to our newlywed status by then. She even got brave enough to give me a haircut, her first. The results was interesting. I tried to keep it hidden under a hard hat.

An Absent Without Leave (AWOL) sailor got careless where a stream named Corn Creek dumps into the main Salmon River at the end of the road below Shoup in mid-July. He was trying to hide from the authorities. He attracted a lot of attention with the major wildfire he created.

I was sent out with a load of supplies for the Corn Creek Fire that first night. I remember looking up into those rocks and cliffs from the end of the road at about midnight. There seemed to be fire as far up as I could see. The firefighters assigned couldn't establish good radio communications with anyone from that deep in the canyon. Indianola District Ranger Wally Mueller asked me to take word back to Salmon that they figured they had a good chance of getting things tied in by daylight. I wasn't so sure.

Windblown burning material from the main fire had created spot fires that covered a lot of extremely rough terrain in very fine "flashy" fuels under Ponderosa Pine timber. That country had historically burned at an eight to fifteen year frequency. Man had kept fire out of it for a lot longer than that through fire control efforts. Maybe God just figured it was time.

The fire moved quite a bit the second day, with crews

in dogged pursuit. I trucked more supplies to the fire line. My youngest brother, Guy, had requested that I pick up a rattlesnake skin for his hat band if I got a chance. I killed two snakes on that run. The problem was that I killed them with a shovel. I'm not fond of snakes. More precisely, I'm afraid of them. By the time I had the snakes dispatched, there wasn't a piece more than two inches long-definitely not enough for a hat band.

About then the Corn Creek fire simply "blew up". I was unloading supplies from my truck and putting them on a helicopter so the pilot could take them to the firefighters who needed them. Things were rapidly falling apart when the pilot took off with a load. He was an African-American with military training and darn good at what he did.

The world was on fire just below the heliport. He had to swoop down over the fire to gain air speed for the helicopter he was flying. The helicopter disappeared into the smoke. We could hear the pilot pushing the power through the engine, trying to gain speed so he could climb. Suddenly everything was quiet. We ran down to see what was going on and met one very wide-eyed pilot climbing back up through the smoke. His engine had simply quit. He somehow managed to make it to a burned-over knoll that just happened to be there when he really needed it. Helicopters don't simply drop like a rock if the engine quits. As long as the main rotor continues to turn they can settle to the ground with some degree of control. The pilot got into my truck and asked to be taken to town. I couldn't get him to say more than a word or two on the three hour trip back to Salmon. Someone told me they met him again a couple of years later. He was dealing blackjack in Reno, Nevada. I don't know if he ever flew again.

Now I could go on the fire as a firefighter instead of a truck driver. I had been upgraded to a sector boss by this time. I was assigned one contingent of fifty SRV firefighters out of Nyssa, Oregon (Garcia was back!), and a second fifty pick ups out of the bars and drunk tanks in Butte, Montana and Spokane, Washington. There was no comparison between crews. The pick ups started to drop out in pretty short order. That's some of the toughest country in the world. It was hell on drunks. The heat, fire, snakes and rocks didn't make it any easier. I still had forty-seven of the original fifty SRV's with me when we left the fire line twenty-seven days later. Three were injured in the control effort-one cut his foot with a Pulaski, another got hit in the head by a rock a buddy threw at a rattlesnake and the third had an allergic reaction to a yellow jacket sting. All of the pick ups were long gone.

I had my first experience with organized Indian fire crews off of the reservations in the southwest on Corn Creek. We had Apache, Hopi, and Zuni crews in the camps I worked. I've worked with a lot of crews along the way and these were some of the best. Some Alaskan crews even showed up before it was all over. They were good workers too, as long as we could keep them away from ants and other intriguing insects and wildlife. Apparently, ants don't do well where there is permafrost, so the Alaskans had never seen them before. They would spend hours watching the busy little critters if we would have let them.

I was also assigned a "contract" crew for as long as they lasted. An innovative individual recognized that federal agencies couldn't maintain a lot of trained fire crews on their limited budgets. He put together some crews with at least some training, and offered to contract them to the agencies when major fires occurred. In this case, the crew boss was an ex-Marine and his crew members were mostly college kids.

In theory, they should have been good. The fire ran over the top of us a couple of times and they went to pieces. They would run every time they heard a tree crown. A running crown fire produces a rather interesting sound similar to a jet engine starting up. Fire burning in ground fuels or in logs or trees next to an evergreen tree dries all of the needles. The tree bursts into flames with a roar. One tree at a time isn't that big of a deal. Now 10,000 trees all in almost instantaneous sequence is a really big deal and a firefighter wants to be certain he is in good standing with God! He also wants to get somewhere safe as fast as he can get there.

We had lots of 10,000 tree events on Corn Creek. I had night operations at first. The fire ran up out of the river breaks on our division north of Long Tom Lookout. It was burning through thick "dog-hair" lodgepole pine. That's tough fuel. Lodgepole pine is one of the first species to return to a burned area. It actually needs heat from something like a fire to help open its cones and disperse seeds. It isn't unusual for too many seeds to take root, creating a dense jungle of small trees that stagnate because of the intense competition for water, nutrients and light. They can grow so thick that it is almost impossible to walk through the resulting "dog-hair lodgepole".

A lot of fingers with unburned islands and numerous spot fires away from the main body of the fire commonly develop in these fuels. Direct line right against the fire is tough and very slow to construct in heavy lodgepole fuels.

We were told to build indirect line, burn out the space between our line and the fire and hold. Theoretically, that sounds like a pretty sound strategy. I've managed to live through such adventures in lodgepole pine fuels enough times to suggest that it's a good way to get killed. The fire isn't impressed by such efforts.

First off, any fire line is tough to build through dense lodgepole. More importantly, you can't get the unburned fuels between your fire line and the actual fire to burn on your terms. Any effort to get a back fire going will fizzle, until things get so dry and hot that I sure don't want to be anywhere in that vicinity. There really isn't a lot of small fuel on the ground to carry a back fire. The crowns are too dense to allow grass to grow under them, and most of the other fuel on the ground burned up in the last fire that created the existing timber stand. There are a lot of log piles scattered around, depending on how long it has been since the last insect epidemic or fire. They burn real hot.

The fire jumps from these burning log piles into the tree crowns whenever the wind picks up on a hot day, and this fire simply runs through the tree crowns with the wind and prevailing slope. I have yet to see an indirect line through lodgepole fuels that will stop a fire that wants to move. The crown fire also throws sparks and fire brands into log piles scattered up to a mile ahead of the main fire, starting numerous spot fires ahead of it. This burning sequence will continue for days on end, until the fire runs out of lodgepole or the weather changes, whichever happens first. Once such a fire really gets rolling it doesn't matter how many dollars (men and equipment) you throw at it. God will prevail.

A new overhead team sent bulldozers to the eastern

portion of Corn Creek Fire to construct line. The day shift constructed a very dusty line that was at least six cat blades (± seventy feet) wide down a major ridge. We were to hold it that night. I had the crews lined out, and staged the bulldozers in strategic points before dark. The temperature remained above eighty degrees after sun down. Then the wind picked up. The fire crowned immediately. A running crown fire coming right at you in the dark is an awesome, terrifying sight. The roar makes shouted conversation almost impossible. Fire brands and sparks were landing all around and on us. Hundreds of spot fires began flaring up across the cat line, threatening to surround us. We never had a chance.

I radioed Crew Bosses and ran down the line, pulling everyone back to the safety zone we had identified on a burned out knoll above us. It was a mad dash for both men and bulldozers through the smoke, dust, fire brands, sparks and confusion to reach that knoll. There was a lot of motivation. I was the last one to walk up the line. The fire was so hot by then that I had to walk along side of the last bulldozer headed up the line, using it as a shield against the heat. The door closed right behind us. We were wearing fire resistant "Nomex" clothing, but that didn't keep sparks from falling down our necks, or burning wrists where gloves and shirt sleeves meet. Most of us received a few small burns on our neck and wrist, and our shirts were riddled with small burned spots from sparks.

We waited on the knoll until the wind died down then spent what was left of the night constructing line along the new edge of the fire.

The overhead team wasn't real happy about our losing what they thought was a secure fire line. They were from the west coast and had never fought a lodgepole fire but should have gotten used to it shortly. Losing fire line became an everyday (and night) adventure for everyone on Corn Creek. The overhead team members rarely ventured far from camp so I doubt that they ever understood what was happening out there on the line.

Night shift is the toughest. A person can only last so long. A sector boss gets briefed at about four p.m., eats, organizes his crews and overhead and goes on the line about six p.m. He supposedly comes off at about six a.m. Assuming he does get off the line as scheduled, he still has to get back to camp, debrief with the day shift, talk with the planning organization concerning what he accomplished during the night and anticipates needing the next night, washes (if there's any water), eats (if there's any food), then tries to get some sleep starting at about ten a.m. Understand that he is in a fire camp with people, and probably trucks, bull dozers and helicopters running around, and generators running. Add a steadily increasing temperature that probably will reach ninety degrees by noon, ants, dust, smoke, flies and other insects. Now try to sleep.

We eventually rolled over to day shift. By this time the fire had burned out of the lodgepole higher country on our end. Now it was burning down slope in cliffs and flashy dry grass and ponderosa pine fuels on the west side of Horse Creek.

A new camp was set up at the mouth of a tributary named Colt Creek, and we moved there. This camp is now in the River of No Return Wilderness. Everything came in by air or was hauled down the main Salmon by jet boat to the mouth of Horse Creek and packed to camp by mule train.

We set up a kitchen, and both the quality and quantity of food improved, providing we made it back to camp when the kitchen was serving. Long distances and fire behavior frequently interfered.

We couldn't get crews into the cliffs and had to work the edges and catch anything that rolled down out of the rocks towards the creek. Sometimes we won, sometimes we lost. My SRV crew was working where the fire had reached Horse Creek one day. It was rattlesnake heaven! Again, most of these folks came from Texas or Mexico, and found our snakes puny by their standards. I had a tough time keeping them from chasing one another with a snake on a shovel. They killed fifty-six rattlesnakes one day in the heavy brush along the creek. Mostly they just shoveled the snakes into the fire and watched them burn.

We faced a problem just below the snakes the next day. A large snag was on fire on the edge of a cliff above the creek when the shift ended. It was obviously about to burn off, and would roll to the creek when it did. We had no night crews in the area because of the hazardous terrain and snakes. I sent my crews back to camp and the division boss (I forget his name), Joe Kinsella (the adjoining sector boss) and I waited for the snag to fall. It came down about ten p.m. We got things under control and started back to camp in the dark without head lamps.

The temperature reached 110 degrees that day. It was still warm after dark. We decided the snakes would have holed up in the shade during the day, but would be out hunting after things cooled off. And now we had to walk for about five miles through them in the dark without lights. We took turns leading, and held a very lively running discussion over whether a snake would bite the first, second or third person who stepped on or near him. We never heard a snake buzz, and no one got struck. We were a little stiff from the tension when we finally tip-toed into camp!

We did have a snake bite on the sector that joined mine on the north. Dick Hickman was sector boss. One of his crew members was climbing up through the cliffs to reach a spot fire that he wanted to work. He raised his head up over a ledge to find himself face to face with a rattlesnake. The snake struck him between the eyes. The victim handled it better than I would have. The first aid provider would have wanted to change my pants as the immediate first aid had it been me. Dick had the victim on a helicopter bound for the hospital in Missoula, Montana within minutes. He survived.

I apparently had a closer call on one of my SRV crews a few days later. We were working our way down slope to tackle a spot fire across the line in dog-hair lodgepole. One of the crew bosses was a huge African-American who claimed to have had previous fire experience on a prison crew. He made a good crew boss. A person would have to be very stupid to give someone that big a bad time. The crew boss was leading the way as we worked down through the lodgepole jungle. He kicked an old rotten log as he passed. He hollered "Jesus, man, get outa heya", then proceeded to rip an impressive swath through those stunted trees as he ran. I was still being impressed by how anyone that big could move that fast through the thick cover when the first yellow jacket stung me behind the ear.

Then we were all ripping our own swath through the trees as we ran. We all picked up some stings. The Hispanic behind me got stung at least six times. He was going into anaphylactic shock by the time we cleared the hornets. We got him onto a helicopter as soon as we could. I understand it was close, but he survived.

We struggled through these adventures for twenty-seven days without a break. We were on what was only one "zone" on this huge fire. It burned down the main Salmon River to the Nez Perce National Forest where personnel from the Forest Service Region One picked it up. It crossed Horse Creek below Colt Creek, where another overhead team took charge. We finally got a brief break in the weather and made enough progress so that some of us could be released.

My crews and I were pretty well worn down by then. I remain very proud of my SRV crews. They were good men. We hiked to the mouth of Horse Creek where jet boats waited to ferry us upriver to the end of the road. Buses met us there to haul us to Salmon.

CHAPTER 10

Sage Creek

Lois and I were happy to be back together again when I got home that night. She even suggested that it was time for another haircut. I was trying to build up courage for that event when I went back to the office the next morning. There had been another fire in a timber sale on Brushy Gulch while I was on Corn Creek. I was sent there to see if we could make an additional timber sale to salvage the trees burned in the fire. The fire was still smoking, although they had it under control. Crews were still assigned to mop up. We could make a salvage sale, so I estimated the timber volume involved and headed back to Salmon with my report.

I stopped at the Indianola Ranger Station when I came to it. The stop was a normal courtesy to see if they had any mail that I could take to town for them. It sped up the information flow and saved the taxpayers postage. Gordon Daniels, the district fire control officer, was pacing the office floor when I stepped in.

A new fire had just been reported on Sage Creek. The patrol plane and lookouts reported that it was really moving. Essentially all of the district personnel were still fighting fire or on assignments elsewhere. Gordon had to stay in the office in case something else started. I said "I'm outa here", and started back to Salmon before he could ask me to take over the fire. A second night at home with Lois sounded attractive.

I didn't get far. I had to pass the new fire on the way to Salmon and saw the smoke billowing up before I even got close to the mouth of Sage Creek. I turned around, drove back to Indianola, and volunteered.

Gordon had been dispatching men to the fire whenever they got in from other assignments. They were pretty well scattered out along the fire line without any coordination or direction. I started to locate them wherever they were along the fire line so we could start an organized effort. Several retardant planes (TBM's) were working the fire hard when I got there. Those converted torpedo bombers were fast and had a short turnaround time from Salmon on this fire. Most of the pilots were old World War II veterans and could really bring the planes down low to drop their load of retardant. The bentonite retardant mixture used then was a thin mud about the consistency of thin pancake batter.

One of the firefighters Gordon had dispatched was working by himself in a pretty hazardous saddle at the head of a draw when I located him. I was trying to get his attention, get him into a safer position and work him in with the rest of the crew when the plane came. We had strict orders to get out of the way any time a plane was making a retardant drop. I hollered as I ran up the steep slope but the noise made by the fire and plane negated my effort.

The retardant drop hit the firefighter head on. A couple of hundred gallons of retardant over a hundred miles per hour obviously packs quite a wallop! The force simply picked the man up off of his feet, carried him about fifty feet through the air, and rolled him down the steep slope. I figured he was dead. The red lump that had been a firefighter a few seconds before was still moving and making some pretty weird sounds when I got to him. I got him sitting up and wiped enough of the retardant off of his face so that he could breath. Miraculously, he hadn't been seriously injured. The first thing he uttered was "My God, I thought that stuff came out as a powder." It does look that way if viewed from a distance.

The fire burned up through some cliffs and mountain mahogany onto a south facing slope covered with dry grass and scattered ponderosa pine. It was moving fast and the afternoon wind was more than enough to keep it ahead of the few men I had. Things were looking pretty bleak when Ranger Orville Engelby and Howard Rackum, his fire management officer, showed up with a crew from Salmon. Damn, but we fought fire that afternoon and night! The wind died down after dark, but the fire maintained a lively pace up the steep slope. We split the crews with Orville's crew working up the west flank and my crew working the east. We could build direct line right against the fire most of the time and began to make good progress.

We had problems whenever we encountered a pitchy stump, log, or fire scar on a standing tree burning close to the fire's edge. All of this area had burned at a fairly frequent interval before man intervened. There were a lot of pitch-filled logs and stumps plus pitch-filled fire scars on live standing trees. Pitch produces a fire so hot that firefighters can't get close to construct line.

A couple of us moved ahead of the line builders, cooling these hot spots with dirt thrown from shovels. It was more effective and faster than trying to build indirect line around the hot spots and dealing with the unburned fuel between the line and the fire.

I was working one of these hot spots about midnight. The fire was burning extremely hot in a pile of old pitchy logs and stumps. When things get hot enough the super-heated air rises rapidly, causing the surrounding unstable air to rush in to fill the vacuum. Depending on atmospheric conditions, serious whirl winds can develop which are especially serious if they start within the fire. Known as "fire whirls," they are full of burning material and hot air, and can spin off in unpredictable directions. I could see the fire whirl coming but just couldn't move that fast. It simply ran right over the top of me as I hurled my body to the side. Fortunately, I turned my face away and closed my eyes as I jumped and rolled. Everything else was covered with fire resistant clothing. It was just a red hot burst then it was gone. Nothing on my person seemed to be actively burning when I stopped rolling and had time to run a check. I now sported some pretty scorched hair and eye brows (maybe the haircut could wait?). I had tried to hold my breath, but my lungs hurt. The fire whirl had spread the fire quite a bit, but we could work the new edge. I rested for a minute before moving on up the line.

By about three a.m. it became obvious that we could cut off the head of the fire if we could just keep moving. We were beat and low on water, but the prospect of winning is a powerful incentive. We met at the lead point shortly after day light. About thirty men had caught an active two hundred acre fire in less than sixteen hours. It felt good. We mopped up some of the worst looking places and re-enforced line. We were desperately low on water, had no food, and were dead tired. Some of the men were so tired and dehydrated they kept trying to throw up. We didn't have any radios with us and could only assume that relief was on the way. A runner came along about ten a.m. with word that we were to be relieved by a Native American crew out of Fort Hall, Idaho.

I stayed on the fire line to show the new crew what needed immediate attention, while Orville moved the rest of the crew down to the road. I could see the buses with the incoming firefighters drive past on the road far below me before Orville got there. They came back past Orville one more time but didn't stop. He assumed they knew where they were going and continued back to Indianola.

It was pushing noon by this time, and getting very hot. The fire wasn't burning much although there was a lot of smoke inside the fire line. I continued to walk the line to make sure something hadn't crossed. Suddenly I couldn't get my breath. My heart was pounding and I went down. I was one scared firefighter. I just lay on the fire line alone. Things slowed down eventually, and I could get some air. I crawled into the shade of a small bush and rested.I never found outwhat happened but assume it was a combination of exhaustion and dehydration. I did scorch my lungs in the fire whirl, which probably helped. That eventually degenerated into a bad cough with lung congestion that slowed me up a bit.

I eventually got up and walked down to the road. The

oncoming crews didn't have directions and were just sitting in the bus. They hadn't thought about going the mile or so back to the ranger station to get direction. I told them where to go (including how to get to the fire), and walked on back to Indianola. I was too tired to eat and had trouble holding the first few sips of water down.

CleeShinderling, a cook I knew from Mackay, gave me his bunk in the cook house. Nothing more registered until about six p.m. when I woke up to hear a lot of activity going on outside. More crews were arriving. You can't imagine my disappointment when I stepped outside and looked up at that huge mushroom shaped cloud of smoke that was the Sage Creek Fire. It had escaped from the day crew. It took us another week to corral Sage Creek.

Then it was back to Salmon and the haircut Lois had promised. I was making preparations for the salvage timber sale on Brushy Gulch the next day when the next fire call came. Lightning had struck a lone snag in the rocks west of Stormy Peak Lookout. No suitable helispot could be located anywhere close, and all of the smokejumpers were out on other fires. A firefighter at Indianola and I were the only ones not already on fires who had gone through the heli-jump training.

It promised to be a pretty rough jump. The helicopter didn't have much room to maneuver in close to the ground and we'd be jumping onto a steep slope. I was closest to the door. I kicked out the fire packs and tools while the pilot hovered. The first pack lodged behind a large rock, as I intended. The second missed, and rolled for a considerable distance. I lucked out on the jump and stayed put when I hit the ground. The pilot made several passes before the other person jumped. I was recovering the wayward pack when he came out. He bounced when he hit the ground and rolled at least as far as the fire pack did. He had some bad scrapes and bruises but we cut down the snag and moped things up. It snowed before we left that one. That was the only heli-jump I ever made on a going fire. Injuries piled up elsewhere and the program, as we used it, was abandoned. It was continued in some areas, with padded clothing, face masks, etc.

The snows continued, ending fire season 1961.

CHAPTER 11

Army Time

I was promoted to Division Boss in the fire organization. An interesting letter from the Selective Service arrived soon after I got off the fires: I was ordered to report for my pre-induction physical for the armed forces.

The Russians blockaded Berlin the day I reported to Boise for the physical. Things looked pretty grim. I was advised to get my affairs in order, since all they could promise was twenty-three days before my draft notice arrived. Actually, the government doesn't move that fast.

An eventful fall progressed into an exciting winter, which, in turn, gave way to a wet spring. Lois and I never knew there was so much money in the world. She was making about $3,200 a year teaching school. My base pay was just over $5,000 a year. When we added her salary to mine, plus fire overtime, we made over $10,000 in our first nine months out of college. That was a lot of money for a couple of poor kids. We had it made.

My draft notice arrived in May, 1962 directing me to report for induction into the Army on July 16, 1962. Forest Supervisor Gene Powers got me a three month deferment because of fire season. For all practical purposes, fire season 1962 never really developed. It rained a lot and normal fire crews handled everything that came up. I never got on a fire but put in a lot of field time, helping to prepare timber sales on various ranger districts.

I left for the Army on October 15, 1962. President Kennedy blockaded Cuba on the day I reported to basic training at Fort Ord, California. I suggested that they really didn't want me in the Army since major international events occurred every time I followed orders. I was apparently the only one to think it might be a problem.

So I got to see Fort Ord, California close up. Basic training at sea level was a breeze for someone who had just spent the summer marking timber sales at around seven thousand feet elevation. It didn't take long to figure out that all a soldier had to do was look busy and the sergeants would give whatever job they had to someone else. I did volunteer to drive trucks, which got me out of a lot of work.

Army personnel asked what I wanted to do when I completed basic training, as if it mattered. I suggested that I might do well in the Combat Engineers, with the idea that I might pick up something there that would be useful in my career with the Forest Service when I was discharged. They found out I could type. Orders arrived sending me to the Army Information School at Fort Slocum, New York to become an U.S. Army Information Specialist. That turned out to be a fancy name for a Stars and Stripes Reporter and, in my case, editor for the "Bullseye" newspaper. It was strictly chance, not choice.

The Bullseye (the soldier's preferred name was the Bullshit), was the official paper for the U.S. Army I Corps (Group), headquartered out of Camp Red Cloud, north of Uijongbu, Korea. I shipped out to Korea in March, 1963.

If I had to be an Army private in Korea, it wasn't a bad job. We covered meetings at Panmunjom, censored the news that might reach the civilian media where it would be sensationalized, wrote home town news releases for all of the soldiers who reported in the area, and put out the Corps newspaper every two weeks.

Having a commanding officer like Lieutenant Colonel John T. Martin helped a lot. We did our job and Colonel Martin took care of us; we screwed up, and we were in trouble. I understood that, and so did my best buddy, Private Laurens Gaskins from Charleston, South Carolina.

A majority of my fellow soldiers spent every spare minute in the many bars and whorehouses in Uijongbu. I was married and held some pretty conservative Christian ideals. I had stood before my God in a little white church in Richfield, Idaho, and promised to love, honor and obey the most beautiful woman in the world until death did us part. I couldn't see any way that a man could maintain that pledge in a whorehouse thousands of miles from the one he loved.

Laurens shared my Christian values and was engaged to be married as soon as he got out of the Army. We did a lot of hunting, hiking and sightseeing in Korea around a demanding job.

Lois taught school in Salmon for the winter then custom mowed alfalfa hay with a new swather her Dad had around Richfield, Idaho that summer and saved all the money she could. Army pay wasn't much, but I sent her as much as I could. I took R&R (Rest and Recreation) leave and met Lois in Japan in August. What an adventure! She even finagled a visa into Korea for ten days although the Army really didn't want her there. Colonel Martin got her a United Nations Correspondent press pass, and she was even able to visit Panmunjom. It's great to see the world with someone you love. Then she was gone. It was a long time until I returned to the States in April, 1964.

I saw two forest fires in Korea. Korea's spring and fall weather can be pretty dry, compared to the feet of rain that fell during the summer monsoons, or the snowy winters. One was a small brush fire on a steep south slope. There were lots of Koreans fighting it, so we stayed out of the way.

I just saw smoke from the other. The North Koreans routinely ambushed our patrols along the Demilitarized Zone (DMZ) although we weren't supposed to be at war at the time. It was their way of producing something to discuss (deny) at the peace talks. They always denied responsibility, and blamed the incidents on our South Korean allies. During one of these battles, tracer bullets from a communist machine gun started a major blaze. We had a strong wind out of the south that day. The fire was heading back into North Korea when I saw the smoke. Reports were that it torched a lot of illegal mine fields and other armament that the Communists weren't supposed to have. Hopefully it burned a lot more.

Next stop was Fort Belvoir, Virginia in April 1964. Lois made that trip with me. She made more money selling ladies underwear in the Post Exchange than I made as a Specialist Fourth Class. We were discharged from active duty on October 16, 1964 and made a very fast trip back to Idaho!

CHAPTER 12

Cobalt

The Forest Service had made no promises but we were able to return to the Salmon National Forest. I worked on a variety of timber sales and timber related projects until an opening developed on the Cobalt Ranger District in December, 1964. Cobalt is forty-eight miles from a paved road, west of Salmon. The ranger station was located four miles above the old copper-cobalt mining town of Cobalt, Idaho. Mining activity had pretty well closed down. A few permanent residents still owned houses in Cobalt and couldn't afford to move, living on welfare and the wildlife they poached. Periodic exploration by other mining companies frequently added some people to the local population.

Gilbert L. (Tommy) Farr was ranger. Tommy, and his wife Jan, were great people to work with, which always helps, especially in those remote "compound" situations. All of the Forest Service families lived in government owned houses, bunk houses and trailers at the ranger station. Some of the accommodations were in pretty poor condition.

No outside work was available for spouses. They couldn't work for the Forest Service because of nepotism regulations. There was no local industry to employ them. Add a few errant dogs that either bit others or tipped over garbage cans, or a kid (or parent) who had problems relating to others in the isolated situation and those close living conditions can produce some challenging social problems. We were lucky.

I spent the winter of 1964-65 living in the bunk house at Cobalt and commuting to Salmon on weekends. Lois patiently waited in Salmon for our first baby to arrive. Jan Marie Pence was born on January 3, 1965.

I apparently hadn't missed all that many major fires while we were gone. The wet spell experienced in 1962 carried through the next two years. Brothers Lew and Carl got on quite a few smoke chaser fires. They made me homesick when they wrote about them while I was in the Army.

I spent 1965 working on timber sale preparation and administration. We picked up a few small fires, but again my primary job wasn't fire control. I did get to go on some.

The first one resulted from a rancher's ditch burning project on the North Fork of the Salmon River.

Later in the summer I took a couple of smoke chasers on a small fire on lower Panther Creek one night that presented an interesting challenge. Lightning had set fire about eighty feet up in the top of a huge ponderosa pine tree. The patrol plane reported the situation correctly, but underestimated the size of the tree. We packed a chain saw on the hike to the fire. The challenge was that the saw had an eighteen-inch bar and the tree was well over forty inches in diameter. I carved away around the entire base and couldn't cut deep enough to make the tree fall. We'd chop out what we could with our Pulaskis and I'd have another go with the saw. Burning debris kept falling on us from above. We posted a lookout to warn us to pull back whenever something fell. It was still tough to get out of the way in time.

A sawyer has a normal tendency to cut down with the saw tip. I ended up over compensating, and cut up instead. The blade finally broke through after a lot of chopping and sawing. Now we had a huge tree setting on a "cone" that was firmly centered over the point formed by my cut. The tree wouldn't budge. We pushed, pried and tried everything we could to make it fall. We could get the tree to rock a bit. It rewarded us with a shower of sparks and burning limbs every time we tried that. Some of the limbs that came crashing down were large enough to kill, so we had to back off. We had some plastic wedges in the saw kit. The tree just ate them. We finally sawed enough wooden wedges to make progress. The tree fell about daylight.

Some pre-suppression fire funds were available, a normal Congressional response every time a serious fire season occurs. Pre-suppression funds are designed either to prevent forest fires or to make it easier to control them when they do occur. They can be effective if used to reduce fuel buildup, or by making it easier to reach a fire.

Helicopters were rapidly replacing us old "ground-pounders" supported by horse and mule pack strings for fire suppression. There are serious limits concerning where a helicopter can land in rugged, timbered country. We reviewed the records and constructed "Helispots" by clearing trees and leveling ground at marked points in high fire occurrence areas.

Little snow fell during the winter of 1965-66. The dry spring that followed promised a challenging fire year. Tommy Farr, promoted to district ranger out of Sun Valley, Idaho, was replaced by Andy Finn. Andy brought a long list of his own stories with him.

I was upgraded to Assistant Ranger, which meant that we moved into a real house instead of a poorly insulated, mouse infested, old trailer that had an oil furnace that kept exploding. We were still depending on an old SPF radio frequency for communications. We had a speaker in the house so we could monitor what was going on when no one was in the office, providing atmospheric conditions allowed messages to reach us.

It sounded like someone was trying to call the ranger station on the Saturday afternoon before Mother's Day. The telephone line was down, so the radio was all we had. I went to the office where I could both transmit and receive. I finally understood that there was a big fire on Dutch Oven Creek, on the adjoining Indianola Ranger District. Joe Denny, the fire dispatcher, wanted us to gather up every one we had and head for the fire.

None of the seasonal fire crews were on yet, and almost everyone else had gone to some more civilized place for the weekend. I was able to get Andy and a couple of other men and we headed for Dutch Oven Creek. The fire was really rolling in last year's dry grass under ponderosa pine timber on a very steep south slope, accompanied by a brisk wind. Personnel from Indianola were just getting some rubber rafts into place so we could cross the main

Salmon River when we arrived. Snow drifts still covered the timbered terrain on the north facing side of Dutch Oven Creek. They would stop the fire's east flank. We started building line up the west flank.

It got dark shortly after we started building fire line through that extremely steep country. Rolling rocks kept raining down on us as the logs that held them in place burned away on the slope above us. About midnight Andy started asking how much farther it was to the top. I kept assuring him, "Just a little farther, Andy, just a bit more." By about three a.m. we were still climbing while building fire line, Andy was still asking, and I was still sticking to the same response. He finally sat down, and demanded "Enough of this bull shit! Just how much farther is it, really?"

"Really, Andy, it isn't much farther," I responded. How did he expect me to know? I'd never been there before either, and it was really dark and smoky.

"How'll we know when we get there?" he asked.

"There'll be a great big sign saying entering Montana," I replied.

Andy was quiet for a couple of seconds (unusual for Andy), then said "Okay." He got up and we went back to building line. We reached the ridge top and the head of the fire about daylight. Actually, we were still a few miles south of Montana.

We had things pretty well pinched off when we started back down to the river that morning. Several crews made up of full time Forest Service employees had arrived from the Beaverhead National Forest to the east in Montana, and from the Payette National Forest farther to the west. They were good hands, and all of them knew how to fight fire. We got a kick out of the folks from the Payette. They all showed up with stickers on their hard hats. They'd picked out the "easy" jobs like "camp officer," "finance chief," "timekeeper," "engine foreman", anything that didn't involve scampering around on the rugged Salmon River Breaks. They'd all been there before, just farther down river.

We needed line workers, not overhead who stayed in camp and watched others climb that steep mountain. They all picked up shovels and Pulaskis, grumbled a bit, and headed up the hill.

The fire wasn't that tough to catch once the wind died down after the fire burned into more moist areas higher on the mountain. We did have a problem when one of the local welfare leaches/poachers tried to steal some supplies out of the camp along the road during the night. Jim Caples was camp officer. He tried to stop the thief, only to have that scum ball try to run over him with his pickup truck. We had the fire wrapped up in a couple of days but the weather continued dry.

CHAPTER 13

Close Contact with a Drunk Driver

We scheduled a big outing with some of my family in Mackay for the opening of fishing season on the first weekend of June, 1966. We made it as far as the Mackay Reservoir where we met a drunk driver head on. He was driving a Dodge pickup truck with a camper and was towing a boat. That combination was more than a match for our Rambler station wagon.

The drunk, who operated a dry cleaning establishment in Pocatello, had an impressive record of drunk driving and related accidents. Witnesses stated that he had gotten too drunk to walk in "Perk's Pool Hall" in Mackay so they helped him into his truck. I'm not sure why any thinking person would do that. The drunk drove past where he was supposed to stop in a local trailer park, side swiped a bridge abutment, crossed the center line and caught us just as we came over a hill.

Drunk driving laws were pretty lax in those days, especially in Custer County, Idaho. He walked away. I broke seventeen bones, including a major concussion and seven ribs that broke in thirteen places, plus a collapsed lung. Mostly it just hurt when I breathed or laughed. I didn't laugh much. Lois cracked some ribs and an elbow and suffered a concussion. We're convinced that our seat belts were all that saved us.Lois threw herself over Baby Jan who was asleep between us. She was able to hold Jan in place for the initial impact before hitting her head on the dash. Jan was thrown clear of the car when it rolled. Thank God the car didn't roll over her. She survived with cuts and bruises.

It looked like a pretty lean fire year for me. However, we had a lot going on, so I was back on the job in ten days. Doctors assured me that I would only regain seventy percent of my lung capacity. They had been so busy trying to keep me alive that it took them over a month to discover that my left knee cap was broken in two places and my left leg had a hairline fracture down from the knee joint.

A lot of fires started but I was able to stay out of it all until early August when reports had a fire building up in heavy lodgepole near Swan Peak, between Salmon and Cobalt. All of our fire personnel had been dispatched to other fires. I had been in the area before. A bark beetle outbreak killed hundreds of acres of older lodgepole in the area about thirty years earlier. Most of these trees were lying on the ground now, with a jungle of younger trees growing through them. We affectionately referred to such piles of logs and trees as a jack straw mess.

I picked up two trail crew members who were at the station and headed for the fire. We were probably ten feet off the ground, walking on down logs when we reached the fire. Dispatch had all of the old TBM's that were available dropping retardant on the fire when we arrived. They knew what the fire's potential was too.

It wasn't that far to the tanker base in Salmon where the TBM's could reload. We received a lot of badly needed air support. The tough part was trying to stay out from under the drops and still fight the fire on the ground.

We'd just get cutting line and a plane would show up. We'd climb back out from the bottom of that pile of logs, scamper out of the way as best we could, they'd drop, we'd climb back down, get started, and another plane would show up.

We were losing ground with that tactic. We could hold our own as long as we could work the fire while the planes dropped. We just lost too much effective ground time getting out of the way. Besides, we couldn't move fast enough every time, so we were getting pretty soaked with retardant anyway. We finally opted to just roll under a log every time a plane came in, and stay on the line. Now we made headway.

There was still a lot of standing dead trees around us. If the bombers didn't knock them down, they burned off at the roots and came down. One of us tried to remain on watch for falling snags at all times. He'd see a snag start to fall and yell "snag!" We'd jump under a log and survived. We successfully kept the fire at less than an acre. They weren't so lucky in 1988 when that whole part of the world went up in flames.

Lois reminded me that I wasn't supposed to be going on fires yet as I iced a swollen knee when I got off the Swan Peak Fire. She asked a lot more questions when I woke her up in the middle of the night, standing in the middle of the bedroom yelling "snag!" I didn't know a nightmare could be that real.

We talked seriously that night. I hadn't realized how much she worried about me when I fought fires. It's easiest that way. She had decided I'd been on too many fires to allow myself to be trapped by the flames. But there were the airplanes, helicopters, snags, rolling rocks and other things that could go wrong. I suggested that I could turn my attention away from fire and concentrate on safer options. Lois, a perceptive lady, knew how much fire meant to me and asked only that I use my head. She also asked that I not tell her about the close calls, and requested that I stop long enough to remember the family before I did anything stupid. She really has me pegged.

I thought we were through discussing the topic when she made one last comment that was to haunt me later, especially on a fire named Mortar Creek. "I know you won't needlessly put yourself in harm's way. What I really worry about is that some idiot will get himself trapped, and you'll feel obligated to try to save him."

There were other fires that followed in 1966. I steadily improved and made it out on some of them before fall storms ended fire season.

Trying to winter at Cobalt didn't make sense. We couldn't maintain dependable communications with the Supervisors Office through the winter so we missed out on a lot of important management decisions that affected us. All but the main access road were closed by snow, and even the major access routes weren't passable a major part of the time. The snow was too deep to work effectively on those steep slopes, even if we could get somewhere to do field work. We loaded the file cabinets and the few personal items we owned in stock trucks and moved to Salmon for the winter.

CHAPTER 14

Assistant Ranger

Our son, Jay, arrived before we left Salmon for Cobalt on April 29,1967. He made the move when he was two weeks old.

A lot of new responsibilities came with being assistant ranger. I still had timber, plus at least some responsibilities for everything else that happens on a ranger district. Stan Feucht, our fire control officer, was responsible for fire organization and control but I still got involved in fires whenever something escaped initial attack.

Some major communication advances were occurring. The Forest Service was moving to high frequency radios which operated pretty much on a "line of sight" basis, making remote places in canyon bottoms or behind mountains difficult to reach. Repeater stations were being installed on mountain tops to solve this problem. The new radios also used new transistor technology that made them much smaller and easy to carry.

We had a particularly interesting fire take off on Big Deer Creek one afternoon in 1967. The fire was reported by a patrol plane but several lookouts could see the building smoke column almost immediately. The fire was on a steep south facing slope with lots of potential. Andy and I plus Joe Denny, fire dispatcher in Salmon, were familiar with the area and involved fuels. We were getting strong winds out of the southwest, fire danger was extreme, and the remote location required helicopter access to get crews there in time to stop it. It looked like major trouble. The regular fire crews were already out on other fires but we had a three-person timber crew at the station.

Joe promised to start rounding up crews, helicopters and supplies but indicated it would be morning before he could furnish us with backup. We had a helicopter at the station when the call came. Andy and I headed for the fire to size things up so we could start plotting our strategy for action as other firefighters came in. Chances of controlling this fire with the initial attack forces we could get on the line that night appeared very slim.

We landed on a relatively flat area just below the fire. I walked around the fire to size up the best control strategy while Andy radioed requests for obvious needs. My review indicated we might be able to contain the fire <u>if</u> we received enough help to get a line around it that first night. A lot of country would burn if we failed.

Andy looked confused when I walked back down to the helispot. "Something went wrong at the station", he said. "Looks like just you and me for the night."

We never found out exactly what happened. The best explanation we got indicated that the helicopter had mechanical problems. Meanwhile a couple of smoke chasers returned to the station from another fire. They got in a fist fight with the timber crew we had ordered to follow us to the fire. Each crew wanted to be first to get on the helicopter. It was too dark for the helicopter to fly by the time the pilot got the mechanical problems solved and the crews got through fighting. Meanwhile, Andy and I had a lot of fire on our hands.

We brought several shovels and Pulaskis and our fire packs with us on our flight in. We looked up at the fire that was racing through the grass and timber above us.

"So what do you think?" Andy asked.

I didn't have anything else planned for the night so I said "You take the left flank and I'll take the right."

We each took a shovel and a Pulaski, tied our lines together at the downhill side of the fire and headed in separate directions. The fire was doing great with the steep slope, wind and flashy fuels under ponderosa pine timber. Pockets with young Douglas-fir under story crowned and burned hot along with piles of logs and other fuels. Andy and I both knew how to fight fire. We built our line as close to the fire as we could so we could simply step into an area that had already burned if something went wrong. We had to step inside on several occasions. The fire would jump our line and come sweeping up behind us with amazing intensity. We'd step into the burned area, walk back down through the smoke to where the fire had gotten away and start building line back up the flank. We fought a lot of it that night. Thank God, the wind died down about midnight.

The work became mechanical-shovel dirt on the hot spots to cool them down; break through the bunch grass and soil with the Pulaski; sweep the loosened material away from the fire with the shovel to form a trail down to mineral soil a couple of feet wide; shovel, chop, shovel, burn out any unburned fuel between us and the fire as we advanced; and repeat the routine over and over on up the mountain. Sometimes we had to back off from the really hot stuff and build indirect line then burn it out so everything was black between our line and the fire. The soil was decomposed granite with few rocks and the line was relatively easy to build. Time went fast, and so did my water. I routinely pack two quart canteens with me. I tried to hold off on drinking that first drink for as long as I could simply to conserve water. I knew I was going to need all I had.

I could see the head of the fire by about an hour before daylight. More importantly, I could hear the "chink, chink, chink" of Andy's Pulaski through all of that smoke and fire. I checked my canteens. One was dry and the other held only a swallow. I think I would have died of dehydration with that last swallow of lukewarm water still in my canteen. I guess just knowing it was there was the important thing, like some sacred object to be worshiped. I kept building line and before long I could make out Andy's dimming headlamp through the smoke. Damn, but we stood a chance.

I don't know for sure how we did it but we tied our lines together at the head of the fire before daylight with probably fifty acres of fire behind us-just a two-man show. We shook hands there in the smoke and dark.

"Do you have any water?" Andy gagged, "I finished mine about an hour ago."

"Just a swallow," I replied. Andy didn't ask for it.

"Have you eaten anything since we started?" Andy inquired.

"No," I replied. We had both been far too busy to think about food.

"I have a can of apricots in my fire pack that sounds great right now," Andy suggested. We both had sorted through our normal C-rations that the Forest Service provided fire fighters. We tossed the crackers and some of the less appetizing meat dishes and fruit and replaced them with more edible things at our own expense. I had a can of apricots too.

"Sounds good," I said. "See you at the fire packs." Andy walked back to check his line and I walked back down mine. Everything looked great. We arrived at our fire packs just as dawn was starting to break, so thirsty and tired that every footstep took unusual concentration.

We sat down together, dug out the prized apricots in heavy syrup, opened them with the C-ration "P-38" can openers, took a good swig of the sweet contents and started to eat the fruit. Then we were both on our knees, throwing up the few morsels that we had gotten down, then moving on to dry heaves that tore at our insides. The dehydration and physical exertion supplemented by the sweet, syrupy contents was just too much.

We had just weathered that experience when we could hear a plane approaching from the east, then the radio came alive. It was Joe Denny. He was so concerned about our fire's potential that he personally flew out on the reconnaissance plane to see how things were going.

"I've ordered five crews (a hundred men)," Joe stated. "They're on their way to the Hot Springs Helispot. Will that be enough?"

"Actually, Dan and I have things pretty well beat down

and lined," Andy responded. "There's still some mop up to do, but one crew should be plenty."

Joe didn't say anything for a while. The plane circled the fire over and over again.

"Damn," Joe radioed. "How did you do that?"

Andy normally wasn't at a loss for words, but his response was simply "It wasn't easy. I don't suppose you have some water you could drop?"

"No, but there will be water with the 'chopper that's right behind me," Joe replied. "I'll turn the extra crews back. You'll be getting twenty men."

We could hear it then, the "whop, whop, whop" of the first helicopter echoing across the canyon far below us.

We just sat there together; a fellowship of two. We had challenged a fire breathing dragon in its lair. We had engaged it in fair combat and we had won. Yet the beast lay behind us, smoldering in its own ruins. It could rise again if the crew didn't clean things up.

I pulled my canteen out of its sheath and handed it to Andy. He shook it to measure the contents, took the half swallow that was his share and handed it back to me to drain. We got up to check the helispot to make sure it was clear so we could direct the helicopter in. We didn't say anything to each other about the night. What we had done was there for everyone to see. We directed the incoming firefighters onto the line, pulled out our sleeping bags and tried to get some rest. It didn't take long to realize that we were too charged with adrenalin for sleep. We mostly drank water and walked along the fire line, checking on the mop up. We didn't want to lose what we had fought so hard to win. Everything held.

CHAPTER 15

A Lot of Fire

A few weeks later, Stan Feucht met Andy and me as we unloaded horses and mules after a week-long pack trip through the Big Horn Crags. He reported that there was a big fire at Reynolds Lake on the Indianola District. They wanted me for Division Boss. I didn't even get a shower after a week in the saddle.

We were getting several organized crews around the country by then. It was obvious that trained crews were much more effective and easier to supervise than pick up crews. Besides, welfare programs were becoming all too lucrative and it was hard to find unemployed men who wanted work. All we had to do was get the organized crews from their home station to the fire line. Transportation alternatives were improving so we could move them effectively. We had some really good organized crews on Reynolds Lake as well as bulldozers and could work some of the line with truck-mounted pumps.

Not a lot of country was that level and developed on the Salmon.

I was working near the lead of the fire when a TBM came in with a load of retardant. It was a good drop spot but he came in low-too low. He took the top out of a Douglas-fir with his landing gear right in front of us. Those were tough planes, obviously made to stay in the air with a lot of holes in them. The pilot jettisoned the retardant load as he pulled up and the plane wobbled away. He apparently made it back to Salmon without farther incident. They had to pry parts of the tree top out of the landing gear and do other repairs before the plane could take off again. I'm sure the pilot had to change his pants before he could go out again anyway.

I got off that fire, went on another pack trip, and was called in early for a fire on Bird Creek on the Salmon District. The Forest Service was starting to depend more on larger four-engine bombers that could carry a lot more fire retardant. They made some impressive and helpful drops on the Bird Creek Fire.

The upper portion of Bird Creek had just been logged. The logging debris (slash) had been piled with plans to burn it after the first snow that fall. They didn't have to wait. Once the fire reached those orange piles (the slash still had dry "orange" needles on it), there was no way we were going to stop it under those conditions. We backed off to a road above the fire, had the bombers lay a blanket retardant above and backfired from the road. It worked like a charm with the retardant putting the damper on any burning material that blew across the road.

Lots of fires burned in northern Idaho and Montana in 1967. Most of our crews had been sent north. We were more than dry enough to burn too, but hadn't had any lightning or careless campers start anything. Andy and I tried to work as close to the station as we could in case something broke.

We took a beating from a lightning storm about mid-August. All we had left on the district was a five person crew that piled slash on timber sales. We sent them to a fire on Hoodoo Ridge on the south end of the Big Horn Crags. They radioed back that things were going pretty well the next morning. Just after noon a follow up message from the Hoodoo Ridge Fire suggested that they were having problems, although the Crew Boss didn't want to admit it. For some reason, most of us are afraid that the world will think less of us if we admit we're in over our head. That was what I heard from that troubled radio transmission. We agreed that I needed to go to Hoodoo Ridge to see what was really going on, while Andy stood by in case something else broke. The crew on the Hoodoo Ridge Fire was more than ready to admit defeat by the time I arrived. Hoodoo Ridge was a fairly typical high elevation fire in Engelmann spruce, white bark pine and sub-alpine fir. We had spot fires scattered all over the mountain, more than the few available people could handle and things were deteriorating in a hurry!

I wasn't going to make much immediate difference, but there's nothing in my personality that prevents me from yelling "help" loud and clear over the radio. The major problem was that there was no help immediately available, especially for a fire on the edge of a primitive area. All trained forces were fighting fires to the north where whole communities and valuable timber stands were burning. We could see several smoke plumes from larger fires on the Nez Perce and Payette National Forests from our lofty perch.

So we did what we could although it obviously wasn't going to be enough. Then Andy picked up a fire in the ponderosa pine/grass slopes and cliffs on lower Panther Creek. We stood a fair chance of burning up the whole Cobalt Ranger District.

Fire management staff in Salmon and Ogden, Utah recognized the manpower shortage and went looking for help. Employees for Kennecott Copper Company were out on strike in Utah. Union negotiations didn't prohibit fire fighting. The strike had been going for some time and some of the workers needed money. We had a labor source. They knew nothing about fire but were willing to learn and got a very basic fire behavior course before they left Utah. They got the rest in a hurry on the fire line on Hoodoo Ridge and lower Panther Creek.

I never cease to be amazed by how little many people living in the Intermountain West actually know about the mountains that surround them. Hoodoo Ridge includes a pretty rugged piece of real estate. It was a full time job just keeping some of the Kennecott people from falling off the mountain. There was no shortage of blisters to treat. I'm not sure what those people did at Kennecott but it obviously had little to do with physical labor. Once we got around all of that they made a reasonably effective fire crew and we started making progress. I'd just think we had it corralled and a spot would take off in a patch of trees somewhere outside the line.

I was patrolling the line one day when I heard someone trying to gobble like a turkey. I went up to see what was going on.

The firefighter said "I didn't know you had wild turkeys around here."

I assured him that we didn't.

He said "Yes you do, there's one wandering around in the ashes right over there."

He was being entertained by one of the biggest blue grouse roosters I have ever seen.

We bantered around "That's a big rooster blue grouse"; "T'is not, it's a turkey," for much longer than my patience normally allows. I finally advised him that I had lived in this type of country all of my life, had killed and eaten a lot of blue grouse, and that was damn sure one tough old blue grouse.

He said, "Well, it stopped every time I gobbled at it."

I said, "Hell, I stopped when I heard you gobble too."

He didn't think my response was nearly as funny as I did. It didn't dawn on me until after I left that he might have been insinuating that I might be a turkey too.

We were getting desperately short of supplies and fire rations were getting old for every meal. We could get trucks to within about two miles of the fire on very primitive roads, but it was tough on equipment and took hours to get the trucks even that close. Most of the air support was fully occupied to the north. That old blue grouse was starting to look pretty tender when I got word that an air drop was en route from McCall. Hoodoo Ridge is about as wide as a timbered knife blade, jutting up above some pretty dissected country. The drop zone we marked at fire camp was less than a hundred feet wide. A miss in that country is as good as a nearly vertical mile.

That old DC-3 airplane came lumbering out of the west and dropped a couple of test chutes to check the wind. Then they got down to some serious cargo dropping. We rarely got to see such marksmanship on the fire line. They hit right on target with fourteen out of the fifteen cargo packets. The fifteenth one is still there. I have no idea what's in it. If anyone is interested, it's about halfway down the slope between Hoodoo Ridge (elevation about 9,5000 feet)the Middle Fork of the Salmon River (elevation about 3,000 feet), which isn't much more than a horizontal mile away. It wasn't worth the risk to send a crew down through the cliffs looking for it.

CHAPTER 16

Clayton District Ranger

We left Cobalt in March, 1969, when I was promoted to District Forest Ranger for the Clayton Ranger District on the Challis National Forest. It was a lot like going home, since I had worked there as a seasonal employee from 1958 through 1960. The District headquarters was located six miles up the main Salmon River from Clayton, Idaho (population 32).

Challis was the nearest metropolis, with a population of around five hundred at that time. About the only way Lois could get "outsiders" to understand just how interesting shopping could be was to tell them that it took her three and a half hours to get to a K-Mart or a shopping mall. The ranger station was located adjacent to a paved highway, a step up from Cobalt.

The "White Cloud Issue" was just coming to a boil on the ranger district in 1969. American Smelting and Refining Company (ASARCO) had discovered molybdenum at the scenic base of Castle Peak in the White Cloud Mountain Range. Close to twenty-five miles of major road construction into a scenic roadless area was required for mine development. Several beautiful meadows along a pristine mountain stream were proposed as settling ponds to support a seven hundred and fifty acre open pit mine. Environmental organizations went wild. I got a crash course in "caught-in-the-middle" the hard way. Everyone wanted to know what was going on. We put a slide program together and I hit the speaker circuit. If the Chamber of Commerce in Payette, Idaho, requested the program, I'd go there and they would try to lynch me because the Forest Service was standing in the way of commercial development. If the Idaho Wildlife Federation wanted a program in Idaho Falls, I went there and they would try to lynch me because the Forest Service was bound by the 1872 Mining Law, and was allowing rape and pillage of valuable resources.

Add normal ranger district responsibilities, and the whole district had to be burning before I could commit much time to fire. Budgets were tough. Our permanent full time staff consisted of me, Bill Millick (district forestry technician) and, eventually, Anita Southwick, business management assistant.

We were responsible for the management of over half a million acres of pristine National Forest. We put on summer crews to help and did what we could. Lois became acting ranger by default more than once when the rest of us were away from the station. Lois had to load the kids in a truck and come pull me out of mud holes on more than one occasion. We had some dedicated seasonal helpers, including Joe Zigler, Connie Cummins, Keith and Leonard Bradshaw, Bart Nordling, Robert Boren, Chuck Ebersole and others, who took on a load far greater than their pay and seasonal status recognized each summer.

Our youngest, Teri, arrived just before Christmas, 1969. Lois had more than a full time job just keeping track of which way three active youngsters and a dog were going in that beautiful setting, especially since the main Salmon River ran through our back yard and Highway 93 ran through the front.

I still hear from some folks who report that they stopped at the ranger station for information only to find that the paid personnel were all out working somewhere on that big district. "We'd start to leave," they report, "Only to have this young lady come out of the house along the road with three little kids. She stopped us to see if she could help us with what we needed." Lois made a good "acting" ranger, without pay.

The nearest high school was in Challis, thirty-six miles away. A school board member called to see if Lois was interested in driving the school bus to Challis in the morning (a 7 a.m. to 5 p.m. job), coaching and teaching, and driving the bus back that night. She would have to travel with the teams to different ball games around central Idaho. That would have been tough under the best of conditions, impossible with small kids and no nearby baby setters.

I am very proud of what we accomplished with that small, very dedicated crew. Fortunately, we were in a fairly wet cycle, and didn't have a lot of fires for the first few years. There was more than enough work to occupy twenty-four hours a day, seven days a week, three hundred and sixty-five days a year without them.

Yet there was fire. My first assignment as Type Two fire boss on a larger fire came during that first summer in Clayton. I was attending a meeting between the district rangers, staff officers and the forest supervisor in Challis when the call came. A fast moving fire was reported in Sand Canyon on the east side of the Lost River Mountain Range just north of Howe, Idaho. The forest supervisor interrupted the meeting to ask who was qualified to take the fire. I was, and I had my fire pack in the back of my truck. A good fire was far more interesting than those meetings. Within minutes I found myself pounding down the highest mountain range in Idaho on a helicopter, en route to the Sand Canyon Fire.

There's a lot more rock in that part of the world than there is water and vegetation. And there was also a lot of sand in Sand Canyon.

That area is in the rain shadow of the Lost River Range, and normally receives less than eight inches of precipitation a year. It must have gotten more than normal that year, because there was quite a bit of dry grass to burn. There really wasn't much at risk as far as resource values were concerned. The fire was burning through some pretty rough country. There was no water so very little livestock grazing had historically occurred in the vicinity, hence the heavy grass cover. The stunted Douglas-fir and limber pine stands offered no timber value.

Some deer wintered on the south exposures where there was mountain mahogany and sagebrush, using snow for water. The sage and mahogany stands were over mature, offering limited value to the deer. Burning would actually rejuvenate these stands and benefit wildlife in the long term. I was under a lot of pressure to order bulldozers, aerial retardant and organized crews, but chose to stick with the local crews on hand. A few acres more burned represented a small cost compared to the major expenses the added personnel and supplies would generate.

Besides, I can think of all too many instances where I have stumbled across an eroding bulldozer trail in the mountains. Sometimes it takes quite a bit of study to determine that what I've found relates to a fire that burned the area several years before. The vegetation has recovered to the point that little evidence remains of the fire except for the cat lines that will remain essentially forever. I wasn't interested in being part of that legacy.

Our strategy was to use hand crews to turn the fire into the rocky timberline above the fire. It took a couple days but it worked. No one tried to second guess us and send the expensive equipment. We saved a lot of taxpayer dollars.

One evening that fall a power line went down just above Sunbeam Dam on the Stanley District. We helped with initial attack since the fire had a lot of potential. The biggest hazard on that fire involved working directly off the major highway that traverses the canyon. Visibility is extremely limited because of the curves and rocky canyon walls. There are a lot of accidents along that road simply because visitors fail to realize they can't see around the next bend, yet feel obligated to drive as fast as they can. Not many people are used to driving winding mountain roads that follow major rivers.

We had a poor fire budget on the Clayton District and were short on equipment. A couple of Stanley back pack pumps looked good to one of my smoke chasers. He hid them in the back of one of our trucks, only to get caught. Normally, I don't make a very good politician, but I managed to give the pumps back and leave without a fight. I instructed the fire fighter to be more careful so he didn't get caught next time.

The forest had a contract helicopter stationed in Challis. I watched it work during fire school one spring, and didn't like what I saw. The ship looked okay, but the pilot really didn't seem to know what he was doing.

A fire took off on Kinnikinik Creek, just above the Clayton Silver Mine one July afternoon. I ordered the helicopter and several crews, and went to the fire. The ten-year old daughter of one of the local miners was standing and crying on the road at the base of the fire when I arrived. I asked her what the problem was. She said she had sprayed charcoal starter fluid on a porcupine and set it on fire just to see what would happen. That started the Kinnikinik Fire.

The crew from the Clayton Silver Mine was already there, but hadn't been able to make much progress. We maintained a Forest Service fire cache at the mine, including shovels, Pulaskis, canteens and rations for just such emergencies. Some of the miners routinely broke into the cache and stole the contents. They were using old shovels from the mine.

The helicopter arrived and I jumped on it to size up the situation from the air and decide where to place the crews that were coming. We never made it to the head of the fire in the air. There appeared to be some real power problems with the helicopter although they may have been pilot induced. We circled the base of the fire several times but couldn't gain altitude. I told the pilot to set back down. I wasn't about to put anyone else in with him, so I sent him back to Challis.

The fire was running towards Poverty Flats and we could

get there in four-wheel drive vehicles. We manned the fire by driving above it and working our way downhill to tie in with crews working up from the bottom. We only violated a couple of standard firefighting orders by that approach. It worked and was a lot safer than using the helicopter.

We had several local ranchers signed up as "per diem guards" who kept Forest Service fire packs and tools at their ranches. We called on them when we needed assistance, and they were only paid while on fires. They would grab their equipment and head for the fire. They appreciated the extra income and knew how to work. Unlike the fire cache at the mine, I can't remember ever having any of the government equipment show up missing at one of the ranches. They were very proud that they could help us. Several of these ranchers were helping on Kinnikinik Creek. We ended up fighting fire all night, and had it pretty well corralled by morning.

Frank Maraffio, one of the ranchers, had an old buckskin horse that followed him anywhere like a dog. Frank would fight fire for a while with the horse munching grass and standing near him. When Frank got thirsty he would take a break, load the horse up with canteens, and they would walk the fire line in the dark so everyone could get a drink. The fire was burning through some cliffs and steep country, but Frank and that old horse never missed a step in the smoke and darkness.

I caught some flak from the Supervisor's Office for not using the helicopter. I told them that it wasn't safe, and that I was putting it off limits to my personnel. That really got the big shots upset, but I wouldn't budge. They brought some helicopter specialists up from the Regional Office in Ogden, Utah. They assured everyone that everything was okay. I advised them that their opinion didn't matter. I had seen that pilot try to work that ship under fire conditions, and it simply wasn't a safe combination. That ship and pilot were a major mishap looking for a place to happen, and it wasn't going to happen with any of my people on board.

Wes Carlson, the forest supervisor, walked past the heliport in Challis en route to the dispatcher's office one afternoon. He found the pilot wrapping electrician tape around the main rotor. The pilot told Wes he noticed a split in the main rotor, but had it taped so it probably wouldn't crack any farther. He was putting the same amount of tape on the other end of the rotor to balance things so the ship would fly smoothly. Wes grounded it, in spite of objections from Ogden. We worked out a trade for another pilot and helicopter from the same company stationed on the neighboring Sawtooth National Forest.

ASARCO's mining proposal in the White Cloud Mountains was becoming extremely political. Actually ASARCO was doing more to pacify the public than the mining laws required, but a couple of adjacent rogue miners with bulldozers kept ripping away mountains which kept the situation front page news with ASARCO getting the blame. The national Congressional Parks and Recreation Committee members needed a junket so they could understand the issues. Wes Carlson and I had to meet several helicopter loads of congressmen and their wives and/or girlfriends in one of the more scenic areas one afternoon. We went in on the ground while they flew.

That fleet of helicopters sure woke up that back country as all of those very important people came thundering in. We were one helicopter short when they arrived. We tracked down the missing ship on the radio. The questionable helicopter we had traded to the Sawtooth had been part of the fleet ferrying the congressmen. It had Idaho's representative Orville Hansen and politicians from other states on board when it crash landed in a meadow in Sawtooth Valley. Fortunately, it came down on fairly level ground and no one was hurt.

CHAPTER 17

Cold Weather Fires

An adjoining ranger was having family problem. Some wives trapped with small children in isolated settings while husbands are off working the back country can get kind of testy at times. This ranger's wife had gone in quest of someone other than small children to talk with and wasn't coming back.

He wasn't taking the news real well. A fire took off on his district one unusually hot October Sunday afternoon. Some hunters had shot a major power line, starting the fire. The ranger was seeking bottle companionship and was in no shape to handle the fire.

The wind was really spreading the fire. Supervisor Carlson called to see if I could take it. Most of the regular fire personnel had either gone back to college or been laid off for the winter. I rounded up everyone I could find at the station. They were sending everyone they could locate, including pick up fire fighters from nearby towns. I even drafted my new brother-in-law. Tom Wittinger had spent the summer working for a local miner for exceptional pay. It was tough work. He found out the hard way why the miner was offering such good wages: he had no intentions of paying anyone. Tom sent Lois' pregnant sister home to live with her folks. Tom advised the miner that he had no place to go and no money to get there, so he was moving in with him until he got paid. I think Tom was the only one on the crew to get paid, and that took a while. At least he could make some money fighting fire along the way.

The wind was screaming and the fire was burning up a lot of country when we arrived. I started crews up both flanks along the steep Salmon River breaks. The wind kept blowing harder after dark, and it looked like we were in for quite a run. I tend to lose track of time when everything is dark, on fire, and I'm working hard. The smoke was too thick to see the sky. Late that night, a long ways up that mountain, I realized that what was falling on me wasn't ash: we'd run head on into a major fall blizzard. We really started to get pounded by the storm. The fire recognized winter and went out. It took until after daylight to get everyone and all the gear rounded up and down to the highway. The temperature was well below freezing and we had close to a foot of snow on the ground when we slid off that one.

We were getting field work pretty much shut down for the winter by December. We had quite a bit of snow on the ground and were experiencing some thirty to forty degree below zero weather. A call came late one night. Campbell's Bar in Clayton was on fire, and the Legion Hall was threatened. Clayton had no fire department: we were it. We loaded up a Pacific Marine Pump we kept warm in the office for just such occasions and headed for Clayton. There wasn't much we could do for the bar so we set out to save the legion hall. Joe Zigler was still working. We set the pump up on the ice on the main Salmon River. Joe got it going, and Bill Millick and I started stringing hose with a lot of good help from the town folks. They didn't have much, but they loved that bar and legion hall and were more than anxious to do everything they could to save them.

Lord, but it was cold! We hit the fire between the two buildings with a hose lay from both sides. It was an effective approach, although some spray would make it through, coating everyone and everything with ice.An old guy had rescued several bottles of black label Jack Daniels whisky from the Bar. He made a relatively unsteady tour around the fire, making sure everyone had a chance to take a swig of that fine whisky every now and then. I wasn't use to drinking anything that expensive on my salary, so it really helped.

We'd hold on to the hose until our hands started freezing to the nozzle, then hand it off to someone else and try to get as close to the fire as we could to thaw out our fingers. The bar was full of booze, paint, ammunition and Lord knows what all. It kept emitting interesting explosions that had us looking for the whisky guy with the free booze to help settle our nerves. We even got the old guy to go down to the river to see if Joe needed some fortification.

At least the rest of us had the fire to keep us warm. Joe was standing alone on the ice at forty below, fending off floating ice that threatened to jam the pump intake hose. Joe sent word back that, instead of a shot of whisky, he needed someone to work on the pump while he started a fire so he wouldn't freeze. A couple of men went down and they got a good blaze going on the ice.

We lost the bar but saved the legion hall. We were losing hose pressure by then. Ice formed inside the hose where the extreme cold worked through the brass connectors. The open space left between hose links was about as thick as a pencil when we quit. The hoses immediately froze solid in place. We had to leave them strung out through Clayton until they finally thawed out in late March.

The local residents wanted to start a rural fire department. The government offered some support, using Army surplus vehicles. Residents had to support the equipment locally once it arrived. They were concerned about wise use of our federal tax revenue (too bad the rest of the world doesn't share that thought?). Losing the bar helped Clayton residents get their fire department going.

CHAPTER 18

New Developments

Idaho Senator Frank Church was trying to find a resolution to the White Cloud issue. He recognized that environmental interests had justified concerns about the 1872 mining law and other resource problems. That made him unpopular locally. The senator's wife and her family owned a hot springs resort named Robinson's Bar for many years before the controversy started. Locals claimed that he was selling out to the environmentalists since he saw an opportunity to make big bucks from the resort.

The family leased the resort to others. A young couple named Joe and Bonnie Leonard managed it and had some other young people helping them. They were a fun group, well suited to run a resort. They specialized in cross country skiing in the winter and kayaking local rivers in the summer. These sports hadn't really taken off yet, but they were doing okay. The biggest problem was that all of the young men had long hair, the ladies sported the bra-less look, and none of them felt that the world revolved around cows and mining. Most local folks viewed them as an invasion of the hippie cult, and found them suspect.

Most of the old pole fences had fallen down on the resort. We sold them a bunch of lodgepole pine for new poles and they set out to reconstruct the fences. The old fence ended up in several huge piles of old, rotten poles in the process. They heard that a major storm was moving in one May afternoon and torched the piles. Someone neglected to realize that strong winds would come ahead of the storm front. The fire took off, burning a couple of hundred acres of sagebrush and Douglas-fir timber. We didn't have our summer fire crews yet, but went to the fire with everyone we had. The crew from Robinson Bar was doing their best to stop the fire although they didn't have the equipment or knowledge. They were a lot of help once we got them equipped and lined out.

The rural fire department from Clayton showed up with their new truck. They were hard workers too, although I could tell we needed to work with them more on identifying what needed attention on a fire that was moving. They put most of their effort into putting out the pole piles that were still burning. All of the vegetation had already burned out around the piles. They presented no hazard and were best left to burn clean as originally intended. We needed to locate the lead of the fire and stop it from spreading.

I stopped to talk to them about it. They made it clear that they knew a lot more about fire than I did, so I left them alone. They weren't in any danger and felt good about what they were doing. It wasn't the time and place to argue with volunteers about what should have been obvious. In time they became efficient fire fighters.

The major power line accessing that part of the world lay in the path of the fire. The vegetation was pretty moist that early in the spring so the fire was putting out some real black smoke. The electricity arced between the wires on the unburned carbon and moisture particles in the smoke cloud, which was fascinating to watch. The fire stopped as soon as the wind died down and the storm arrived.

Alan Getty, a pretty outspoken local resident, worked on the new fire department. Alan always made a point of letting me know what he thought, whether I was interested or not. Actually, I enjoyed his visits since at least I knew where he stood. It was tough to tell with some of the others. Alan came into the office right after the Robinson Bar Fire.

"You gonna charge them hippie kids for damages," he demanded.

I assumed he was seeking punishment for them. I responded that I considered it an accident, and would charge them with failing to control a fire. That amounted to a twenty-five dollar fine versus a couple of thousand dollars.

"Good," Alan responded. "That's how you handled that miner when his daughter set the porcupine on fire, and that rancher who let his ditch burning fire get away a couple of years back. I don't see any difference here." I was proud of Alan.

That fall a transient family moved into an old cabin on Treon Creek that had gone through provisions in the 1872 mining law that passed it into private ownership. Dad was on a workman's compensation disability free ride. I never saw Mom. The two teen-age boys made their living on welfare, poaching wildlife and stealing camping equipment from unsuspecting tourists. The only thing I ever caught them at was dumping private garbage into public garbage cans in a campground and starting a fire. I wish I could have done more.

Dad was burning garbage in the late fall of 1971 when the fire got away from them and headed up a hill. Burning conditions weren't all that bad so I wasn't that worried but apparently the fire dispatch folks in Challis were. They ordered a retardant plane from Boise without telling me. We had the fire out and I was questioning the family when the retardant ship arrived. I'm impressed that the pilot even found us since the burned area wasn't that big and there wasn't any smoke to guide him. I had him dump his retardant load on the couple of acres that had burned and return to Boise.

It's hard to be too tough on someone who screwed up when the folks backing you just wasted several thousand dollars. He paid the customary twenty-five dollar fine.

CHAPTER 19

Yankee Fork Ranger District

The White Cloud controversy came to a head when Congress established the Sawtooth National Recreation Area (SNRA) in 1972. The legislation paid ASARCO a hefty sum for not pursuing their mining proposal. The Forest Service established some badly needed new mining regulations at the same time.

Environmental groups were elated but most of them lived elsewhere. We had to live with the local folks, and they wanted the mine and related opportunity to obtain good paying work.

There also was the opportunity to make a bundle from inflated real estate values. The local economy was in bad shape, primarily because ninety-five percent of the county was federally owned. It's tough to operate a county government with such a minimum tax base. In addition, there were no good paying jobs available so most young people with much potential left, their parents wanted them closer to home, and the mine had offered that prospect.

The government eliminated the option for the ASARCO mine. The locals are good, extremely conservative people who were unhappy with the government which was symbolized locally by the U.S. Forest Service. We weren't real popular.

A new challenge was presented by the extensive internal reorganization necessary to accommodate the new political boundaries established by the Sawtooth National Recreation Area. The White Cloud Mountains and the corridor along the main Salmon River went into the SNRA. I inherited essentially everything between the SNRA and the Middle Fork of the Salmon River, which was named the Yankee Fork Ranger District of the Challis National Forest. That's about 750,000 acres of some of the most beautiful wild land in the world. Quite a bit of it is in the River of No Return Wilderness now.

New management challenges came with the expanded district boundaries. Our timber management activities expanded and we picked up Dick Foster, a forester, to take on timber, recreation and a lot of other duties. Dick and his wife Susan were a welcome addition to our group. Joyce Rovetto joined us to help with the business end of things and we eventually added Bob Rameriz to help with the timber load.

I was exploring the new ranger district boundaries in the Cape Horn area north of Stanley with my family and in-laws one Sunday afternoon when we spotted a fire on the ridge west of Lolo Creek Campground, near Collie Lake. Roughneck Lookout radioed it in right after we did. The fire was burning in dense lodgepole timber, high enough on the ridge to pick up strong winds. It was really starting to move. I requested ground crews, smokejumpers and aerial retardant, grabbed a truck at the Cape Horn Guard Station and went to Lolo Creek. A helicopter picked me up there.

The fire had a lot of potential. I was directing retardant drops, ground crews and jumpers from the helicopter. Usually the retardant planes have priority over other traffic. It was obvious that we weren't going to make much headway until we got people on the ground. We had already received several retardant drops when the jumper plane from Idaho City arrived. I held the retardant planes off and sent the jumpers in. The air was getting very congested with two retardant planes, the jumper plane and a pretty slow and puny helicopter all circling the fire. There was an open ridge across the Marsh Creek canyon from the fire. Those rocks offered a good vantage point, and we couldn't get run over by an airplane if we landed there.

We didn't figure we'd have any problem landing on a flat area on the ridge. The smoke column off of the fire indicated a very strong west wind. The pilot brought the helicopter in pointed to the west. We were almost on the ground when the pilot exclaimed "Shit!" I've never heard another exclamation over the helicopter intercom that deserved more attention. Even I could tell that we were in trouble. We slammed into the ground in the hardest helicopter landing I've ever made. The landing skids were bent a bit, but everything else held together, although my nerves were a bit frazzled. Don't take anything for granted in the mountains. The wind was screaming out of the west over the fire. And then the mountain topography took over. The wind started to tumble as it topped the main ridge and encountered the junction of three major canyons. We landed with a strong tail wind! Thankfully the helicopter could still fly.

We got the jumpers on the ground and the retardant planes went back to work. The jumpers cut a helispot, I flew into it, and we started ferrying in the ground crews. We weren't all that far from Boise and Idaho City, got good retardant support and stopped the fire that night. It was the only major blaze in the northwest at that time. I was amazed at all of the new innovations in fire control that were offered from various sources. We really didn't need them, once we got on the ground and the retardant kept coming. I could have spent a lot of money on that fire but I pay taxes too.

CHAPTER 20

Jeanotte Creek

Another fire call came on Labor Day weekend. The Payette National Forest had a major fire going on Jeanotte Creek on the South Fork of the Salmon River. They needed a Division Boss.

A Type I Fire Overhead Team under Karl Spellman was assigned. Andy Finn, my old boss and firefighting buddy from our time at Cobalt, was Line Boss. It was good to work with him again. I drove most of the night to get to McCall to tie in with the team. We drove to the South Fork at the end of the road beyond Warren, Idaho. The fire was just across the river.

I was assigned the north flank of the fire. We ferried in the Redding Hot Shots, the Plumas Hot Shots, and some local crews by helicopter and were supported by another crew of jumpers and hot shots from Prineville, Oregon who either jumped or came in via helicopters near the head of the fire further to the west. The McCall jumpers and local crews worked the south flank under another Division Boss.

The fire was burning up out of the break country in ponderosa pine/bunchgrass fuels. It would build up in the fuels below us and make a run at the ridge where we were constructing and holding line. We constructed a lot of indirect line in granitic soils and it burned out well. We'd build line like crazy until we tied into a rock cliff or other anchor, then burn out the fuel between the line and the fire. It worked great.

Then we'd move above that cliff and continue up the slope until we could burn out again. The fire was still going strong when it got dark. We were supposed to leave the line in time for the helicopter to get us back across the river to camp. The fire didn't agree with that plan since it kept running at our line and we were making too much progress to stop.

The fire made a major run at our line shortly after dark. The crews I had knew how to fight fire. We had a crew burn out everything they could between the fire and our line, while the rest of us held the line and patrolled for spot fires behind us. We had a lot of anxious moments, but the line held.

We were a tired, hungry group of firefighters when the fire died down about midnight. Not everyone had brought their head lamps with them. We paired those with lights with those without and climbed down through the cliffs. It worked well.

We'd worked hard and it had paid off. The fire lit up the world to our right. Again, a group of skilled firefighters had met a major fire head on and had made impressive progress. The fellowship of fire became very real. Crew members called back and forth through the darkness, and every now and then some of them would come up with a song or chant that kept moral high. At one point, I looked back up through the rocks, smoke and darkness. As far as I could see were the lights of fire fighters behind me. We had met the enemy on its terms, and won the day's battle. Life was good.

The biggest problem now was how to cross the South Fork of the Salmon River in the middle of the night with about a hundred firefighters to reach camp. The river was too deep and swift to wade, and there was no way to swim fully equipped firefighters across in the dark, even if they could swim.

One of the hot shots had radio contact with another crew in camp. Some huge ponderosa pine and Douglas-fir grew along the river. The hot shot radioed his buddy to see if they could locate two big trees opposite each other on both sides of the river. With any luck, we could saw them down at about the same time so they both fell together across the river. Hopefully the current would jam them together and we would have a bridge of sorts. I didn't have any better suggestions. The other crew was waiting when we reached the river. We sawed down the selected trees. They fell at almost the same time, and wedged together in the middle of the river just as we hoped. We limbed them up enough to get through. The hot shots routinely carried some rope to help get through cliffs if necessary. We strung it across as a guard rail. The trees tips had sunk into the water until they were about knee deep at the point where they came together. We draped our boots around our necks and crossed bare footed. We reached the east side without mishap.

A catering service had arrived to feed everyone, so we ate a great, if late, supper of warmed up chicken, rolled out our sleeping bags in an abandoned apple orchard, and called it quits for what was left of a very short night. Morning comes early on the fire line. It was back up the mountain shortly after day light. All of our line had held, but there was a lot of fire still above us. The fire had burned under a thicket of twenty to twenty-five foot high ponderosa pine reproduction along the ridge top during the night. It hadn't crowned in the tree tops, but had dried the foliage until it wouldn't take much heat and wind to make that happen. We really didn't need a crown fire running at our line that afternoon in that type of fuel. I called in retardant drops to coat the crowns. It worked well. We had another long day. We were helicoptered onto the line in the morning, fought fire all day, hiked down the line to the river after dark, waded across on our improvised bridge, ate a nice chicken dinner and crawled into our sleeping bags.

Most of the gear belonging to the Prineville crews was still at a drop spot high on the mountain. Rather than have to haul them up the mountain in the morning and then have them climb back down in the evening, we just flew them in some more supplies and let them stay up on top during the night.

The rest of us would eat a pancake and scrambled egg breakfast, helicopter to the head of the fire, fight fire until things cooled off after dark each day, hike back down and cross the river on our improvised bridge to camp. We'd eat a chicken dinner and crawl into our sleeping bags for the rest of the night. We were tired enough that I didn't realize that we had eaten pancakes and eggs for four breakfasts and fried chicken for five nights in a row until Andy Finn told me they had fired the caterer. He didn't have anything except chicken for supper, pancakes and eggs for breakfast, and white bread and baloney for lunch. A new kitchen was set up to feed us.

I've never been on a fire with more black bears that seemed to be attracted to the fire for some reason. I almost ran head on into a big one in the dark just as we were leaving the line one night. It reared up to look me over. It was close enough that I could tell it was a boy bear. I resisted the urge to run like hell, took a couple of steps backwards and held my ground. The bear grunted a couple of times, then dropped down to all fours, said "woof," and ambled off.

It was tough to walk the line in the dark by myself after that. The main camp was set up in an abandoned apple orchard next to an old illegal cabin that had been built under the guise of the mining laws. The apples still weren't ripe, but the bears didn't care. They loved the green apples. The cooks got up one morning to find a young bear on the serving tables. They yelled at it and it spun and ran the full length of the tables before it jumped off and ran away. Running was the least of its problems. It left a very smelly string of second hand green apples that squirted out of its backside as it ran down the tables. Some of us didn't have a big appetite for breakfast that morning. Fortunately the breakfast menu didn't include apple sauce. We did have an apple in our sack lunch, but at least it hadn't been run through a bear.

I worked with the hot shots and jumpers for the entire fire. They were great folks. Jim Grant was in charge of the Prineville crews. Charlie Caldwell headed the Redding Hotshots. The Plumas Hotshot crew leader's first name was Joe, but I forget the rest. I told them what I wanted and got out of the way.

Affirmative action and some other amenities had reached the fire line. The crews came with the first females I had encountered on the line. They confirmed what I had always assumed. Some men fit in well on the fire line, some don't. Women work the same way. I remain concerned that if things really went to hell, some women will lack the physical strength and endurance to run to safety in that rugged country. But then there are men who wouldn't make it either. The biggest change they brought was that we became a lot more careful where we went to the bathroom.

A female member of one of the crews made a point of stressing that even that courtesy didn't make any difference to her. She'd just drop her drawers and go whenever and wherever she had the urge. Her "presentation" appeared designed as a challenge to her male counterparts. She was obviously out to make a point, one that most of us didn't consider that important. That lasted until I came along the line and found a pile of human feces lying in the middle of the fire line. I called the crew boss over and asked what was going on. He called a crew member over to load the mess on a shovel and bury it away from the line. I told the crew boss to discuss the issue of taking care of human waste with his crew so that didn't happen again. He was talking to the lady when I went on down the hill. It must have worked since I didn't step in anything that unpleasant for the rest of the fire.

I also discovered a real luxury on Jeanotte Creek. We finally got in before dark one evening. The camp included a portable shower! It seemed like the fire line was getting over-civilized. Still, a hot shower sure felt good.

The Occupational Safety and Health Agency (OSHA) had taken an interest in the potential health hazard that smoke presented to firefighters. A firefighter breathes in a lot of smoke as part of the job. OSHA wanted to determine how much smoke was enough and a testing crew was following the Redding Hot Shots. They waited around camp all day until we came in off of the line. Then they had crew members exhale into some tubes attached to equipment that determined the amount of carbon dioxide and monoxide included. They repeated the process in the morning to determine how fast the body recovers after inhaling a lot of smoke. They had established some baseline standards and were trying to determine how long a person could work on the fire line without exceeding their standards.

I was interested since I inhaled a lot of smoke too. They were quite open to questions, although I had a few problems with some answers. The biggest one I had involved crew members who both fought fires and smoked. I don't remember the baseline figures. The amazing point was that a non-smoker started at a given point and wasn't supposed to exceed another one. The base line for smokers actually started at a higher point than was deemed safe for a non-smoker and went up from there. The OSHA data was used to decide that firefighters should not spend more than twelve hours on the fire line each day, with a break needed after fourteen days. That remains the standard today. It's a tough standard to hold to, especially when a fire is burning hot and a crew has to stay on the line longer to keep from losing their fire line.

It rained after several days on Jeonotte Creek Fire, with snow in the high country. We had to wait two days before the weather broke long enough for us to get the wet Prineville crews and their gear off the mountain.

I enjoyed working with an organized overhead team.

All of the members had regular Forest Service jobs, but were committed to go on fires when called. I was especially impressed by the fellowship involved and how well they worked together. They invited me to join them on a full time basis but my boss had reservations. I needed to tie down a lot of loose ends on the new district before we could let that happen. Besides, there wasn't a landing field anywhere in the Clayton vicinity with lights, so I couldn't always catch a plane when they needed me. I had to decline the offer, but asked that they keep me in mind. I'd like to join them whenever the opportunity arose.

CHAPTER 21

Snake Creek

I got called out as an unassigned Division Boss in 1973. North Idaho and Montana were burning. I flew to Missoula to be assigned. Firefighters carry "red cards" that lists their qualifications. Staff personnel meeting incoming overhead at the jumper base found out that I also qualified as Safety Officer. Since there was a bigger shortage of Safety Officers than Divisions Bosses, they decided to send me out as Safety Officer. They didn't know which fire I would be going to, so I was sent to a motel for the night. Lois' sister Marti lives in Missoula so I called to let her know that I was in town. She dropped down to see me later that evening. A very busty lady in skimpy night clothes answered the door. I can imagine what my sister-in-law thought. I had left for a fire some time before and the lady had been assigned to my old room.

I was assigned to the Snake Creek Fire adjacent to the Selway-Bitterroot Wilderness Area as Safety Officer. We traveled to the fire in a school bus for most of the night.

The Forest Service was starting to figure out the importance of natural fire in maintenance of specific vegetation and the animals that depend on certain habitats. A test area had been established in the White Cap drainage in the Selway-Bitterroot Wilderness to evaluate the option of allowing lightning caused fires to burn their natural course in isolated areas. A large lightning caused fire was burning in the White Cap Fire Management Area. Everything went as planned until the fire crossed the management area boundary. The Snake Creek Fire was that portion outside the test area.

Safety officers are supposed to check all operations to make certain line and camp activities are conducted safely. They emphasize the results of their observations at all briefings and in operation plans, and make recommendations to avoid accidents. In my experience, most safety officers sat around camp drinking coffee and looking official.

A Safety Officer's job can and should be very challenging and rewarding. They are free to go anywhere and do about anything they want. I'd go wherever it sounded like there might be problems. If it looked like the fire was going to make a run at a section of line, I'd go there. If someone was sling-loading supplies in or out of an area with a helicopter, I'd go there. If someone was hurt, I'd get there as fast as I could to help with evacuation. I'd evaluate the accident and determine how to keep it from happening again.

Snake Creek includes some very steep country. The fire was simply doing what fires have done in that type of fuel and topography forever. We were still in the wilderness area and no real resource values were at risk. The danger was that it could build up a massive front and the wind would drive it outside of the wilderness boundary where it could create major problem. It had a long ways to go. We were supposed to stop it. It cost the taxpayers a lot to do that, but would have cost a lot more if we had been trying to put out the whole fire.

A base camp (Incident Command Center) was set up at the end of a road where it was easier to truck in supplies and personnel. We were helicoptered to a spike camp on top of a mountain right next to the fire. All of our supplies had to be flown in. They tried one food drop by parachute from a fixed wing aircraft. It missed the camp by half a mile horizontally, and about that much vertically. We could look across a very deep canyon and see black bears feeding on the remains within a few days. They flew everything we needed in via helicopter after that.

The Job Corps program was just starting out. Most of the enrollees had never been off of city pavement and knew nothing about the mountains. We had one fire crew from the Trapper Creek Job Corps Center on the line. They were good kids but didn't know a lot about what they were doing in that steep terrain with rolling rocks, fire and other hazards.

One of the Job Corp kids got hit by a rock and was injured badly enough to require evacuation. In interviewing him, and watching the others work, I found that this crew did a lot of horsing around. They'd just get to working up the slope and one of them would yell "Rock," indicating that a rock was rolling down the hill. Everyone would jump out of the way and look for the rock. They'd just get back to work and the same thing would happen. Sometimes there was a rock. Usually there wasn't. It was very disruptive behavior. The kid who got hit just got tired of jumping for nothing, didn't duck, and got hit when there really was a rock coming.

I tried to get them to quit the horse play and other little things that could get people hurt. They didn't listen very well since clowning around was simply too ingrained in their way of life. They were also getting a lot of blisters and sprained ankles. Those kids just could not work safely in steep terrain. I had to recommend that they be moved to a safer work environment if we wanted to keep them on the fire line.

We also had a Job Corps camp crew. They were younger kids, real characters who helped around camp. The most exciting thing that had ever happened to them appeared to involve either off-loading a helicopter bringing in supplies or loading it back up with refuse and other material leaving the fire.

They were afraid to leave the camp, having seen the bears eating the wayward food drop across the canyon, and there obviously were snakes around (or why would they have named the fire Snake Creek?). Each evening they'd go out to where the flatter mountain top dropped off at the edge of camp and start baiting each other. Pretty soon they'd scare themselves and come running back into camp.

A commissary was set up in camp so we could get new socks and other necessities and the charges were deducted from our wages. Some free amenities were available. Usually it's just candy bars, gum, pop, and maybe some fruit. Most of the Job Corps camp crew smoked although they were all under eighteen. For some reason cigarettes and even some cigars were included as freebies on this fire. The kids cleaned them out.

I came back to camp early one night late in the assignment. The logistics section had a huge unsorted pile of candy bars, fruit, cigarettes, cigars, tooth brushes, etc. in the middle of the commissary. They'd obviously searched the kids.

The kids could not resist free access to such goodies and stashed all sorts of treasures in their sleeping bags. Some of the stuff wasn't free and had simply been stolen from the commissary when it wasn't manned at night. In some instances, the commissary crew didn't think there was enough room left in the sleeping bags for a person.

"I just know they're going to hit us again tonight," the commissary officer told me in the chow line that evening.

"Just play along when I come up later," I said.

The Job Corps kids were camped right next to the commissary. Several of them had already crawled into their sleeping bags when I came up just before dark.

"Where do you want the snake," I asked the commissary officer.

"Oh, you got my snake!" he responded, obviously pleased. "Just give him to me. I want to put him in with the supplies."

"But he'll bite you," I argued.

"No he won't," he responded. "I'll know he's in there, and I'll just work around him. We've been losing a lot of stuff and I want him to bite whoever's been rippin' us off."

One of the job corps kids sat up in his sleeping bag. "You got a snake, man?" he inquired.

"Sure do," I responded, pointing to the commissary officer. "He said he wanted a snake, so I got him a snake."

"Where you got him, man?" the kid inquired.

"Right here in this plastic canteen," I said, pointing to an empty canteen on my belt.

"How'd you get him in there, man?" he asked.

"Oh, I just held his neck down with a stick until I got his head poked in. The rest just naturally followed," I claimed.

"You kiddin' me, man," he responded.

"No I'm not," I stated. "You wanna hold him?" I offered to give him the canteen.

"No way man!" he cried, and ducked down into his sleeping bag to avoid the offending canteen.

I gave the canteen to the commissary officer, who promptly made a big show of taking the lid off, commenting "Wow-he's a big 'un," and stuffing the canteen with the "snake" in with the supplies behind his makeshift desk.

"They never touched a thing!" he told me at breakfast the next morning. We had no more troubles there.

We had a lot of Indian crews working out of the camp. When things started to cool down later in the fire, a lot of blue and Franklin (fool hen) grouse started to work along the edge of the fire. They'd dust in the ashes to get rid of external parasites and hunt for toasted grasshoppers in burned areas. The Indians loved them. I requested that they not throw rocks or tools at the grouse since they might hit one another. They didn't listen any better than the Job Corps kids, but then I have killed a few fool hens for supper myself.

Two of the Job Corps kids were best of buddies. One was a red headed white kid about five feet tall. The other was a black kid who was over six feet tall but couldn't have weighed over a hundred and thirty pounds. He was mostly arms and legs. They looked like the old cartoon characters, Mutt and Jeff, and were a fun couple to watch.

This was obviously their first big adventure away from the streets in some big city. It was a whole new world to them.

The tall kid had just crawled into his sleeping bag one evening when the short kid came running up.

"They's a man down there got him a budd!" Short exclaimed, all out of breath.

"You jivin' me man," Tall responded.

"Nope, I'm not neither," Short repeated. "They's a man down there got him a budd".

"What's he a doin' with him, man," Tall inquired.

"He got him on a stick, man," he responded.

"You jivin' me man."

"Nope, I'm not neither."

"What's he a doin' with him, man."

"Got him on a stick, and he's a cookin' him, man."

"What's he a gonna do with him, man?" By this time Tall was out of the sleeping bag, pulling on his pants and boots.

"Sayz he's a gonna eat 'im, man."

And then they were both running down the slope to watch this man who was cooking a poached grouse he had skewered on a stick. The pigeons were probably never safe in Central Park following that experience.

It took a couple of weeks to corral Snake Creek before I was sent back to Missoula for re-assignment. A lot of fires were being demobilized. There were hundreds of people passing through Missoula, either going home or to other fires. A major camp was established at the Missoula

Smokejumper Base to coordinate movement.

I was assigned as safety officer there. It was a pretty boring job, one that didn't seem necessary. Our son was scheduled to head for his first day in the first grade in a couple of days. Kindergarten was not available in Clayton, so it was a big deal for the family. I convinced the lady managing personnel assignments that I really was surplus and needed to go home. She was an understanding person. I barely got home in time to see the young man off to school.

We had lots of slash and other debris to burn on the American Creek Timber Sale that fall. We'd had a long, hot fall, and were concerned about getting the slash burned. We finally got a favorable weather forecast about mid-October. Saturday was going to be nice, it was going to snow a couple of inches Sunday, Monday was going to be fair, then a real major fall storm was moving in for the rest of the week.

I had a deer hunt set up with the family and some friends, but Dick Foster assured me he could handle it. Dick is one heck of a worker, as were a couple of guys and a gal he had working for him. I would have bet that they couldn't get everything on fire in one day but they did.

It cooled down a bit Sunday but the wind blew like crazy. No snow fell. We spent most of the day trying to keep the fires where they were supposed to be with mixed luck. The major fall storm turned out to be ten days of seventy degree weather with strong afternoon winds.

We burned up all the slash and then some. Still, we kept most of the fire within the cutting units. We had to call for help, which was most embarrassing.

CHAPTER 22

Nevada

Our children attended a one-room school house in Clayton through the fourth grade. Then it was off on the thirty-six mile bus ride from the ranger station to Challis. The school bus came by the ranger station shortly after seven a.m. and returned about five p.m. The younger kids got off at Clayton and played games until school started. The older ones stayed on the bus until it reached Challis. Lois substitute taught at the Clayton Grade School when needed, and would have liked to pick up employment outside of the home if such opportunities were reasonably available.

Jan was in the sixth grade and Jay was in fourth grade. They had to make the long trip to Challis. Teri had been in the first kindergarten class offered in Clayton. She was going to be the only student in first grade. She obviously needed to be challenged by other students so we opted to have her continue on to Challis where there were more students.

Never play poker with Jan, our eldest daughter. She learned all the tricks by the time she was in the sixth grade on those long bus rides. She also beat up the local bully, who was two years older and much bigger than she was, on the bus one morning. He slapped her little sister. Jan came over the seat like the tigress she was and blacked both of the bully's eyes and bloodied his nose before the bus driver could pull her off. Tough country makes tough kids. Jan was to the point that she wanted to participate in school activities such as band and sports. She couldn't do that and catch the bus for the long ride home. It was also obvious that the powers that be were determined not to upgrade me in place although the work load justified it. With all of the kids in school, Lois needed to be putting her talents to more productive use as well. I started putting in for transfers.

We were offered District Ranger for the Ruby Mountain Ranger District headquartered in Wells, Nevada. It represented a promotion for me, a chance for the kids to be more active in school and other programs, and employment opportunities for Lois. We froze all of the house plants and most of the fruit Lois had canned over the years in the U-Haul on that cold January move in 1977.

A lot of people have asked what I did that was so bad that we got sent to some place as isolated as Wells, Nevada. They obviously had never spent much time at Cobalt or Clayton. Wells had a population of about twelve hundred people. We were downtown! We were even able to join the Presbyterian Church.

Wells is a very friendly town and we had some new country to explore. There's a lot to Nevada that most folks never see. Nevada is best traveled at night, if all you get to see is what's next to the interstate highway. There's some beautiful country there for people willing to get off the beaten path.

Besides, the Bureau of Land Management was trying to get some badly needed management going on lands they administer that had been long neglected. Ranchers are very protective of doing things their way. The BLM got to wear the black hat for a change instead of the U.S. Forest Service. It was neat to be the good guys, or at least not the worst guys, for a change.

A lot of the ranchers and miners in that part of the world still think that anything required by the government is part of a Communist plot. Most of the federal range has been seriously impacted by overgrazing. The more progressive ranchers were willing to admit it and wanted to improve management. The ranchers who felt threatened by improved management efforts launched what became known as the sagebrush rebellion, designed to see all public lands placed in private ownership where they could rape and pillage to their short-term monetary benefit.

The best part about the new move was that I had several assistants who were able to handle a great deal of the routine work. My primary responsibilities were management, supervision, administration and the politics involved in trying to improve resource management and community relations. The district represented a combination of the old Wells and Lamoille Ranger Districts. We had some tough problems relating to the combination, and old work habits that die hard. Actually the Regional Forester advised me when I accepted the assignment that his major priority for the district was in getting the personnel problems straightened out so we could proceed with improving resource management. Some good people were involved; they just had problems working together productively. There were also some employees who had performance problems and tended to misuse both time and government property. The job promised to be an interesting challenge.

CHAPTER 23

A Fire Team Opportunity

Another good feature with the new ranger district was that I now had the time, and access to an airport with lights, so I could join one of the organized fire overhead teams. Karl Spellman was still interested so I joined his team as division boss.

1977 was an interesting fire year. There hadn't been a lot of snow that winter or moisture that spring. Our first assignment was to a fire in the Bridger Mountains just north of Wyoming's Sweetwater River. I flew in a single engine Cessna from Wells to Big Piney, Wyoming where I would meet the rest of the team.

The flight to Big Piney was extremely rough since we continually had to dodge thunderstorms. These were the notorious "dry" lightning storms that plague the west and causes fires since they contain a lot of lightning but little rain. The air was so rough that the stall buzzer entertained us for most of the flight as we bounced from updraft to downdraft. We had to make several passes at the Big Piney airport to chase off a herd of antelope so we could land.

The pilot was anxious to get away since another thunderstorm was approaching from the west. He dropped me off then headed west to skirt the storm and return to Elko before dark if he could. There was no one at the airport. I searched the area and couldn't find a pay phone. I was stranded at the airport all alone, not counting the herd of antelope and a few thousand mosquitoes. I wasn't even sure where the town was from the field.

I sat against the wall of an old hanger and watched the storms rumble by. I paced the strip. Since this was my first official assignment with the crew, I worried that the pilot had dumped me off at the wrong place. I slapped lots of mosquitoes. I had just decided to start walking down the road to see what I could find when another Cessna came in to drop off a fire team member from Utah. He wasn't overly talkative and about as uninformed as I was. At least I could share the mosquitoes.

We were happy when the rest of the crew showed up in an old DC-3 just before dark. They explained that they had to wait for the thunderstorms that we flew through to get out of the area before they could fly. Some of them were still air sick from the rough ride.

An overhead team gets briefed by the "line officer" (local administrator in charge of managing the land that's on fire) when they take over a fire. This briefing includes everything local personnel know about the fire, present and predicted weather, fuels, crews and equipment already committed, resource values at stake, etc.

No real resource values appeared to be at risk with this fire. The fire was just burning up some un-merchantable trees and brush in some rocks. Although fairly large, it was about to run up against the elevational timber line, which consisted of a couple of miles of solid rock that extended to the top of the mountain range.

A Type One Overhead Team amounts to a pretty expensive group of people. They know how to run a fire or other incident, but it costs the taxpayers a bundle every time they are activated. I suggested that maybe it would be best to just let the fire burn, as the area has historically, since nothing of any significant value was threatened. The ranger was more than ready for that question.

An agency must have an approved fire management plan before a fire like that one can be allowed to burn. The Forest Service had started the public involvement process necessary to initiate a fire management plan for this area. The proposal recognized the important role that fire played in maintaining a healthy forest environment and how both vegetative condition and wildlife values would benefit from allowing periodic burning similar to what had historically occurred.

The reason the plan had never been completed was simple: local ranchers were afraid of fire. They called their congressman, demanding that he stop the government from coming up with anything as ridiculous as a policy that would allow some fires to burn during specific burning conditions. The ranger produced a letter from that congressman to the Secretary of Agriculture, that was then sent to the Chief of the Forest Service in Washington so demanding, and that was the end of that discussion. Politics interfere with a lot of needed land management that might benefit the environment. We spent lots of taxpayer's dollars on a fire that really didn't need fought. Fortunately, nobody got hurt.

That is some neat country. The fire was typical of those burning in high elevation timber stands, with lots of spot fires interspersed with unburned timber. The area that burned should be some outstanding wildlife habitat now. I'm sure it's a favorite spot for local hunters.

We got the Sweetwater Fire corralled in a week or so by simply running it into the rocks at timberline, them mop-ping up. At least we thought we did. We released all of the crews and loaded onto a bus for the long trip back to Big Piney for the agency debriefing and trip home.

Just as the driver closed the door, someone looked back up on the mountain and said "What's that?"

We all looked. Way up on that mountain, just outside the fire line, a lone tree was crowning out. We had combed the entire area for spots, but had managed to miss one. Fortunately, it wasn't on my division, so I didn't need to be too embarrassed.

We unloaded, formed a fire fighting crew out of the overhead, and headed up the mountain. The tree was burned up when we got there. We put out all of the fire we could find, checked again for spots, and were a couple of hours late getting to Big Piney.

CHAPTER 24

Dried Bananas

We hadn't been home long when we were called to a fire on Idaho's Caribou National Forest; another high elevation fire typically consisting of lots of spot fires. The crews arrived ahead of other supplies. Notably missing was food and mosquito repellant. All we could locate was some freeze dried fruit and nuts that came in with an air drop. I was assigned as line boss for initial action, which was the first night shift. I loaded up with water and a big bag of freeze dried bananas and headed down the fire line.

The white bark pine at this elevation had been infected by an insect outbreak in the 1930's that killed a lot of trees. Piles of logs lying around subalpine fir carried most of the fire. They would burn hot, crown out some live trees, and throw a bunch of sparks into the next bunch of logs. Vegetation between log piles was still green and not burning. Trying to round up all of the scattered spot fires was the challenge.

I encountered a crew working a particularly hot spot about midnight. I stopped to help them out. One young fellow seemed especially excited. He was still dancing around when we got things cooled down.

"This sure is some fire," he commented.

I agreed. They all tend to be unique in their own way.

"It's not nearly as hot as some of those Salmon River Breaks fires though," he commented, wanting to show how experienced he was. "Have you ever been on one of them?"

"The first fire I was ever on, back in 1957, was on the Salmon River Breaks," I responded.

That quieted him down. I ate some dried bananas with the crew and made sure they all understood what they were doing. The kid hadn't said anything. He finally spoke up just as I was getting ready to move on up the line.

"1957," he said. "That was two years before I was born." He sure knew how to hurt a thirty-eight year old. I used my shovel as a cane as I hobbled around the fire for the rest of the night.

Actually, that wasn't my only problem. Those freeze dried bananas were pretty tasty and I was hungry. I ate a good sized bag of them since they were the only food I had, and then drank quite a bit of water. I have never had such gas. I think they expanded to about two bushels of bananas inside my stomach!

I was still grumbling and emitting some other obnoxious sounds when I got back to camp about daylight. The rest of the overhead crew found the situation hilarious. Andy Finn even had a completed "CA-1" accident report form filled out for me when I returned to camp the next morning. He claimed that I had blown my "bung-hole" off, and he had witnessed the event. He wasn't too far off. They were a bunch of characters, and could make relatively unpleasant adventures like that one seem hilarious-part of the fellowship of fire.

CHAPTER 25

Scar Face

We finally got food, mosquito dope to ward off an impressive swarm of those pesky critters, and caught the fire. We were being debriefed at the Forest Supervisor Office in Pocatello when the next call came. California was burning.

We boarded a plane, bound for the Big Sur Fire somewhere in southern California through a very dark night. We never made it. They diverted us in the air to Lakeview, Oregon. We were bused to a new fire, Scar Face, on the Modoc National Forest in northeastern California.

The Scar Face Fire was burning through some lava rock breaks just southwest of Alturas, California. The country was the scene of a series of major battles between a small band of Modoc Indians, who objected to having their homeland taken over by the whites, and the US Army back in the 1870's. It's interesting country, and most of it was on fire when we arrived.

Lynn Sprague, another division boss, and I were taken to the origin, near Happy Camp Guard Station. The fire had spread north and west from there, and was going great.

"Pence has the east flank, Sprague has the west," they said.

"Where do our divisions end?" I was naive enough to ask.

"We have no idea," they responded, and drove off in a cloud of dust. They were right. A lot of California was on fire. Things were so smoky they really couldn't tell just how big each fire was. The agencies had a limited amount of infrared equipment that can make that determination but it was fully occupied in more populated areas. The convection columns from several of the fires, including Scar Face, showed up on the weather satellites.

Scar Face was novel in that it was the first fire I had ever been on where the basic terrain was flat enough that I could drive most of the line in a four-wheel drive pickup. That was good, since we sure had a lot of country to cover. There were lots of lava rock outcrops, but I could drive around them. It was big timber country with quite a bit of logging. We used bull dozers to construct most of the fire line and burned out from the cat line and the main fire. We held the lines and picked up spot fires with hand crews.

The country was so dry that year we could get a four-wheel drive pickup stuck by high centering it in the fire line's dust and had to change air filters in the vehicles daily.

I have rarely found animals that have been burned over in a forest fire, contrary to what the "Bambi" folks would have us believe. Wild animals evolved with fire, and usually get out of the way, but a band of domestic sheep on Scar Face weren't that smart. The fire burned over the whole band while they milled around in a dry meadow. A lot of the sheep died on the spot. The rest stood around with their heads down, smelling a lot like burned wool. The owner tried to get them on trailers to move them to a veterinarian, or at least somewhere it wasn't all black and smoky. Most of them died before he could get them out. We sent a bull dozer back to bury the bodies when things cooled down.

We had essentially no air support. Greater values were at risk on other fires. Mostly it was just too smoky for air personnel to see what they were doing, even if the aircraft were available. We couldn't even get a helicopter so we could scout the fire.

A lot of this part of the Modoc had been logged several years before. Young ponderosa pine trees were growing great on most of the old cutting units. Most of these young trees (fifteen to twenty feet high) had just been thinned. Thinning involves cutting the less dominant trees to get a more ideal spacing between the dominant trees so they maximize growth rate and reach a commercial size in less time. The cut cull trees were laying scattered around on the ground with dry orange needles still attached. They burned very well. I hate to think how many thousands of dollars had been spent thinning the stands that were burned. There was no stopping the fire when it reached the areas littered with orange slash.

We were trying to catch a spot that had jumped the cat line and was running through some of that thinning slash when we got our only retardant drop. Heat from the fire was building up enough that the super-heated air suddenly formed a convection column that towered over us like a huge mushroom cloud. That smoke column was pulling enough air into it so that the sky where we were cleared sufficiently for a low flying aircraft to see the ground. A retardant plane apparently just happened to be passing by. He came screaming down, dropped his load without direction, and went on his way. The only thing he accomplished by dropping a load of retardant on a fire that hot was to waste a couple of thousand dollars of taxpayer's money. Had I known he was there and could have established radio contact I might have been able to direct him so he could have at least helped open an escape route for us. We were in real trouble at the time.

I was driving a new rental brown Dodge four-wheel drive pickup truck. It was a great truck. The fire was rapidly surrounding us and we had to make a run for it. I loaded the crew in the pickup and we took off. Most of the crew members were stacked on top of each other in the back of the truck. We went flying down an old logging road overgrown with vegetation just as the fire closed in from both sides. There were flames everywhere. I managed to stay on the primitive road by referencing the outline of darker tree trunks that stood out from the flames on both sides. One of the smaller burning trees crashed down on the road ahead of us. We bounced over it and somehow shot out of the flames. We scorched the paint on the new truck-better the truck than our hides.

We were trying to burn out from a fire line in a meadow a few days later. There were some strange structures standing around. "Those look like goose nest stands that I've seen on wildlife refuges," I commented to one of the local firefighters who was working beside me.

"They are," he replied. "This is really a lake. I caught bass about where you're standing a few years ago. The lake just dried up during the drought."

We had some interesting crew problems on Scar Face. One involved an all-woman crew. I think they were called the Morning Stars. A lot of folks were still having problems figuring out just what "integration" meant on the fire line, hence an all-girl crew. They worked fine as firefighters, and undoubtedly would have worked just as well if they had simply been placed in "mixed" crews.

They had been assigned night shift for several nights. They eventually wore down, just like anyone would. We gave them a day off before shifting them to day crew. They radioed in the first afternoon in desperate need of drinking water. The weather was dreadfully hot. Firefighters need a lot of drinking water. Everyone had been directed to pack at least half a gallon of water, plus a can of juice for lunch. We couldn't figure out why they needed water when everyone else had packed enough. Unfortunately, the local water was highly mineralized and really didn't taste that good. Water gets very warm in the canteens firefighters carry, and ends up about the temperature of lukewarm tea. It beats going thirsty.

That crew had been packing hot cocoa in their canteens instead of water while they were on night shift. They got away with it during the cooler nights but the cocoa soured in the hot daytime temperatures. We had to send a special truck with water, plus new canteens.

I was walking down the fire line with a crew from Minnesota a couple of days later. I wanted to show them a special assignment that needed attention. I stepped off a rock ledge, only to note a huge rattlesnake all coiled up right where my foot was going to land. I swear that I turned in mid-air and landed back on the ledge, well behind the firefighters who had been following me.

The Minnesotans had never seen a rattlesnake. They set out to kill the snake while I was busy figuring out wheth-er I needed to change my pants. It crawled under some rocks, and I called them off. A lot more people get bitten while they are trying to catch a rattler than by accident. I showed them their assignment then went on down the line. They showed up in camp that night with a new snake skin; they just couldn't resist.

I had several "accidents" on my crews. One seemed a bit unusual. Since most of the firefighters were men, a "lady" on one of the camp crews figured out that she could supplement her fire wages by applying the oldest profession at night.

We didn't have individual back pack tents for crews then. Sleeping bags were spread in the open without much consideration for privacy. The supply unit normally stocked rolls of plastic sheeting so crews could make "lean-tos" if it rained. We had portable toilets set up around the camp. The lady in question entertained male clients in one of the toilets. I'm not sure how that worked. There's hardly enough room for one person to sit down and do his thing in one of those cramped plastic commodes. Recreational activities seemed impossible to me. The toilets aren't all that stable when set up on ground that isn't all that level. I'm sure the smell wasn't what a person normally associ-ates with a boudoir, although I am not an expert on the subject.

The accident report claimed that one of her clients got a little too excited and tipped the toilet over in a moment of intense passion. I suspect the actual story was that a couple of his buddies waited until they thought everything was about on schedule and tipped the structure over as a joke. The toilets didn't get pumped out too often so this one was very full of human waste as well as the chemical preservatives. I'm glad I wasn't the medic on duty. The amorous couple got covered by all of the contents. Apparently the chemicals used in those toilets were not intended to be in contact with human skin, especially in sensitive areas. I doubt that some of the other contents were all that compatible with anything. They were treated for chemical burns to the eyes and other sensitive body parts. I don't think anyone broke anything.

Scar Face ended up being "zoned." Fires can eventually get too big for one overhead team to handle so another team eventually came in to manage the south and west flanks.

I got a day off after twenty-one days on the fire line so I called Lois and the kids from a pay phone in Alturas, California where we went to wash clothes and drink beer. They still thought I was on the Big Sur Fire since that was the destination I gave them the last time we had communicated. Big Sur was getting a lot more publicity, since it was closer to bigger cities and the media. They made a point of catching the news each evening, hoping to see dad on TV.

We had our zone under control after about a month. I wanted one more look at my division before we left. That way I could note any problem areas on a map that I would leave for the person replacing me for the mop-up phase.

We were able to borrow a helicopter from the other zone. It was a Hughes which is a pretty fast ship. It's interesting to look down on the history of a fire from the air. I could see the fire line lost and what worked and could remember all of the heartbreak that went into losing a long section of line we had worked so hard to hold. There's also the good feeling about the tight spots where we held. It's all written on the ground. Lost fire line shows up well as a brown, almost white, scar through the black terrain in a fire. We lost a lot of line on Scar Face.

We hadn't gone far when a red light started blinking on the helicopter's control panel. It kept flashing "low on fuel, low on fuel!" It's the sort of thing that gets my attention.

"Ah-h, what sort of problem does that indicate," I asked, trying to act nonchalant, and very brave. "No big deal," the pilot replied. "We still have about twenty minutes of fuel left."

It really would take more than twenty minutes to see what I needed to see. We were already about ten minutes out, and "about twenty minutes" seemed like a pretty nebulous figure to me! I was ready to go back immediately, but held out as long as I could. We cut a lot of corners on the fire line. The map I handed over to my replacement was based mostly on what I had recorded while walking and driving the line since the air review wasn't that complete.

One scene will remain with me forever. We were nearing the end of my division. A huge convection column loomed ahead of us on the other zone. Flames leapt high into the column as combustible gases ignited. Below us we could see several bulldozers, and the yellow shirts of hundreds of firefighters walking single file, all working their way across the broken terrain towards that column of smoke. It would take time, and most likely a break in the weather, but they would eventually catch that part of the fire as well.

I'm not sure how much fuel remained when we got back to the helispot, but I was very happy to have solid ground under my feet. We were replaced by a Type Two Overhead Team that afternoon.

We returned to Lakeview, Oregon for some rest. Several of our crew members were sick: We'd been on the fire line for too long. After a day they sent us home for a longer break. I was sure that we'd be heading back to California in a few days. There were just too many fires burning there. It stormed before that call came.

CHAPTER 26

Nr. Nebo and Corral Creek

We had a real cold snap that winter. It came at the right time to freeze the leaves on live oak brush in Utah and Colorado. Live oak burns very well without help. It didn't need freeze-dried leaves on the plants to help carry fire.

Several members of a hot shot crew were trapped and killed on a fire in live oak in Colorado in 1978. They just took too many chances and the fire caught them. About that time, our team was assigned to the Mt. Nebo Fire near Nephi, Utah. It was burning in fuels and terrain almost identical to those that trapped the fire fighters in Colorado.

The Mt. Nebo Fire spread aggressively the first day. Most wildlife can usually get out of the way of fires. The black-tailed jack rabbit population was at a real high point on Mt. Nebo. The fire ran over the top of a lot of them. There were dead rabbits scattered all over the burned terrain, with a lot of scorched bunnies still running around the country. Karl Spellman, normal Fire Boss for our team, was sick, so Andy Finn took over as Fire Boss. I went on as Line Boss in his place. It didn't take us long to control the fire and return home.

The fire season didn't look all that bad in our area. We set up a family salmon fishing trip to the Washington coast with brother Lew's family for early August. Lois and I had been unable to take a summer vacation since we got out of the Army in 1964 and the family thought it was about time. We had a fishing boat and motel all reserved on the coast in Washington.A fire broke just on the California side of the border north of Reno, Nevada about ten days before the vacation was scheduled. Lois and the kids issued strict orders that I needed to get the fire out fast or the family was going fishing without me!

That was a big order. The fire was burning in Jeffery pine and white fir which burns a lot like a mixture of ponderosa pine and sub-alpine fir. It moved fast with a lot of spot fires and kept us busy. I was assigned as Line Boss and had some excellent crews.

Charlie Caldwell and his Redding Hotshots were back. I kept pushing-doing everything I could to meet the deadline for the fishing trip.

I normally don't use much aerial retardant on a big, fast moving fire. It's expensive, does little to stop a fire that's moving and crews have to move out of the way for the drop, which slows effective line construction. It normally takes a lot of time for the plane to drop, return to its base, fill and return. The bomber crew is also exposed to significant risks when dropping retardant from a hundred feet over a raging wildfire during a hot, windy afternoon.

Stead Air Force Base wasn't far from the fire with a

turnaround time for the retardant plane of just minutes. I decided to take advantage of this opportunity in an effort to speed up control. Two problems surfaced with the first few drops. First off, the pilot thought he knew more about fire control objectives than I did, and dropped where he wanted instead of where I requested.

The Redding Hot Shots were trying to hold the fire at a low saddle in the head of a canyon, and I was with them. We had a good line established, and were burning out from the line. Our biggest fear was that sparks and fire brands would start spot fires behind us. I wanted the retardant drops behind us on the east side of the saddle to prevent spot fires. The pilot ignored my directions and kept dropping on the west side, trying to hit our backfire. We wanted that stuff to burn, and he kept trying to put it out.

Fortunately, he kept missing our backfire. He was wasting our time and the taxpayer's money so I radioed that we should stop the drops. It was an excellent drop spot, and the "air attack" people on the fire assured me they would see that it was done right. I agreed with John Madden, the air attack officer, to give the pilot one more chance. I'll admit that I made some less than complimentary remarks about the pilot over the radio, some question his mother's relationship with his father at the time he was conceived during my conversation with Madden.

Madden radioed that the bomber pilot was coming in for the drop. I pulled the crew out of the saddle and we got behind some large Jeffery pine trees on the slope above the requested drop site. We were a couple of hundred yards from where he was supposed to drop. We could hear that big bomber coming, but something sounded wrong. I peeked around the left side of the tree since he was supposed to be dropping from about our elevation, to our left. All I could see was his right wing. He was coming right at us!

A lot of things all come together in a person's mind at times like that. It was obvious that the pilot had taken exception to my remarks. He came into that saddle looking for yellow fire shirts, not fire, and found us. I hit the mike on my radio, trying to call him off. Too late!

It's a miracle that someone wasn't killed. The retardant tore the tops and branches out of the trees. They rained down on us just ahead of the retardant. I hugged that huge tree. Firefighters were swearing and ducking all around me. The retardant just sucked in around that tree and caught me from both sides. "Slurp" it said as it covered everything with a couple of inches of red slime. I had to rake the underside of my nose with my shirt sleeve before I could even breathe.

Branches and retardant were still dropping as the prop wash from that huge plane swept over us. Several very soaked fire packs and a chain saw were picked up by the impact of the retardant drop and were sent rolling down the slope towards the fire. Things were so slick that we couldn't run after them.

Even my glasses were coated and I had to take them off before I could see. Everything was red from the dye in the retardant. A bunch of confused, and very red, firefighters were milling around, trying to figure out where their packs and tools were. We all had retardant dripping off our noses.

My radio was coated, but I triggered it anyway, and called Madden. He answered.

"What's going on down there?" he inquired. "All I

could hear was cussing and swearing over your radio."

I told him. If I hadn't questioned that pilot's ancestry before, I did then. I demanded that the S.O.B. be grounded and asked for a full investigation before he could fly again. I emphasized that it was a miracle that he hadn't killed someone. On the bright side, he demonstrated that he could hit a target if he was mad enough.

I don't know the outcome of the investigation. He didn't get another shot at us on that fire. I hope he never flew again. We caught the fire at the saddle.

I made it home for the vacation with minutes to spare. The family was sitting in the car when I reached the house. Lois did mention that she and the kids were getting tired of having to do all of the packing and other work while I was off playing with fire. She mentioned it several times, and still brings it up on special occasions.

We had one interesting district fire that fall near Harrison Pass on my ranger district. One of the ranchers in Ruby Valley reported the fire. Her husband and most of the local ranchers were members of a volunteer fire department supervised by Nevada Department of Forestry. They were already on their way to the fire. The fire was burning up a lot of country when I arrived with crews. A confused gentleman with an interesting story met us at the base of the fire. Part of my responsibility was to find out how the fire started, so I listened. He was a member of a large party of hunters from Las Vegas. Deer hunting season opened in a few days. They got a nice camp all set up the night before, drank a fair amount of whiskey, and the rest of the crew went fishing at Ruby Marshes that morning. This old boy decided to stay behind to rest and guard things (his version). I diagnosed a major hangover.

He'd just gone for a little walk when he looked back at camp and noticed a lot of smoke. He ran back to camp and found everything on fire. He tried to move a truck out of the way, but the ammunition in one of the tents that was on fire started to go off so he just ran. The tents, guns, one tire on their truck and the rest of their gear went up in smoke. He didn't have the slightest idea how the fire started. It was obvious they had a large campfire built in an area surrounded by very dry vegetation. A spark from the campfire took care of the rest. They had a very expensive, if short, hunting trip.

CHAPTER 27

Corral Creek

The fire season started early in 1979. Karl Spellman, Fire Boss on our Type One Team, had been suffering from serious health problems for some time and died that winter. Some major revision in fire teams resulted and I was assigned to a fire team with Gordon Stevens as Fire Boss.

Our first fire in 1979 was the Corral Creek Fire in the Uinta Mountains near Vernal, Utah. I had a touch of flu when I got there and didn't get off to a very good start with the team. I had to work in the planning section until I could hold food and water down. Things had cooled down some by the time I reached the line.

We had water drops from helicopters going one afternoon. The helicopter had a tank attached to its body, rather than using the large "buckets" normally slung on cables under the ships. I hadn't seen such equipment work before, and was amazed at the improved accuracy.

The fire had burned up steep slopes then through some

flatter "mesa" terrain at the top. We had a crew working just under the rim on one division. The helicopter came in for a drop right over the top of them. They dropped lower in the draw they were working in to clear the drop area. The ship hit a tree with a main rotor, almost flipped, then pulled back over the crew in an effort to get away from the mountain. It made it. We'd have lost quite a few firefighters if it hadn't.

We were hiking back to camp one evening when we heard a lot of very excited radio traffic. They normally don't encounter a lot of real crises so it doesn't take much to set them off. It took us a while to understand that they had found out that one of the camp crew members had "crabs" (body lice). They just knew everyone using the camp would get the crabs, and the world was going to end. We couldn't keep away from crabs while I was in Korea, so it wasn't a big deal to me. A lot of the firefighters with me had been to Vietnam, where they had the same problem. I said all we needed was a little G.I. louse powder. The Vietnam vets said that wasn't used any more since it was mostly DDT. A new shampoo/body wash was current state of the art. They finally located some of the soap that killed the little beggars, so things settled back down.

The Uinta National Forest doesn't have many major fires. We got the fire controlled and mopped up until we would normally turn it back to the forest for final patrol and mop up. Keeping a Type One Team activated is an expensive proposition, and was quite unnecessary in this instance. A Type One Team normally gains control on a fire, puts out anything burning within a given distance from the fire line (usually a hundred yards, depending on fuels and fire danger), then turns the fire back to local crews.

They were afraid to release us since they didn't know what they needed to do next. Most big timber fires continue to smolder in logs, stumps, dead roots, etc. back inside the line until winter snows put them out.

They weren't about to let us go as long as they could see smoke anywhere on that fire. There was a fairly large block of unburned timber in a very rocky area way out in the middle of the fire surrounded by a mile or so of totally burned country. There was essentially no chance anything would escape from there to cause future trouble.

That unburned area was still throwing up smoke long after everything else was out. The fire was smoldering around in decaying organic material on the ground and most of the trees would have survived if left alone. It would have been essentially impossible and very costly for us to put all of that fire out, and would have taken all summer.

We preferred to leave that fairly large island of unburned timber so it could serve as a seed source to help new trees become established on the area burned. It also helped the aesthetics, and the older aged trees would meet habitat objectives for some wildlife species. The Uinta folks continued to make it clear that they weren't about to release us as long as they could see that much smoke. We were costing the taxpayers a lot of money.

I took the Sawtooth Hot Shot crew with all the backfire fusees we could carry and set out to see if we could settle that issue one morning. We had a lot of fun setting everything on fire. We produced a lot of smoke and fire, burned everything that would burn, the smoke died down and we went home.

CHAPTER 28

Mortar Creek

The summer continued hot and dry with numerous fires burning in central Idaho on the Salmon, Boise, Challis and Payette National Forests. Our next call was to the Mortar Creek Fire on the Challis.

Mortar Creek was started by a couple of fishermen on a horse pack trip along the Middle Fork of the Salmon River. The fire started in the bottom of the canyon, jumped the river, and was now burning up the steep slopes on both sides. It was over twenty miles to the nearest road with a flat area large enough to set up a decent fire camp. Advance members of the overhead team were setting up a base camp at Bruce Meadows, a grass landing strip on the Boise National Forest, the closest point reasonably accessible by roads and fix wing aircraft when I arrived. I made a quick reconnaissance of the fire on the plane as I flew in.

The Mortar Creek Fire was throwing up a lot of smoke and burning a long way from camp. We had a quick strategy meeting and I suggested there was still quite a bit of daylight left. We already had some good hot shot crews and several helicopters. I volunteered to take the crews we had and catch what we could that night. The best terrain for a night crew was on the west side of the river. No one had a better suggestion. I flew in on the first helicopter with part of a crew, located a reasonable helispot and we hit the line. We got everyone in before dark and took on a lot of fire. Those were outstanding crews. A Division Boss usually has Sector Bosses, Line Scouts, and others to help keep things organized. Those specialties hadn't arrived on the fire yet, so we were on our own.

Fortunately, most hot shot Crew Bosses qualify as Division Boss with several crew members having other qualifications as well, which helped. I tried to keep things coordinated with the crews spread out, working in safe locations, and kept things scouted out so I could place them in new assignments as they completed each segment of line. That was tough duty with a lot of walking in the smoke and dark through rugged terrain. I would walk along the edge of the fire ahead of each crew, flag the needed line location with engineering ribbon, then hike back to discuss what was needed. They'd get going and I'd work ahead of the next crew.

It didn't take long for the night to pass. On a flat map, we probably constructed only three miles of fire line but on the ground it was closer to ten miles of good line, all burned out and secure for the moment.

Rolling rocks and falling snags pestered us all night. A local crew was working up out of the river bottom on the north flank. With the exception of a small section between our helispot and the line the local crew was constructing, we secured the entire fire west of the river by daylight. I was waiting with a crew for a helicopter to take us back to camp at about eight the next morning on a high ridge overlooking the Middle Fork. We were a very tired bunch but our spirits were high because we knew we had done the best possible job under very difficult circumstances. A small black cloud passed directly over us just before the helicopters arrived to pick us up. A single bolt of lightning streaked down to rip a big fir tree apart within a hundred feet of us. We were showered with bark and limbs, but no rain fell. Maybe God was telling us something. He had my attention. I've always had a lot of respect for lightning: it scares me to death!

An accident investigation team showed up in camp just after I gave up trying to sleep that afternoon. A helitack member from another forest had been killed on the Ship Island Fire just downstream the day before. Our camp had the best access to helicopters and the Ship Island Fire for the investigation team. I knew most of the investigators so I got to sit in on their briefing before getting my crews back on the line that evening.

All line personnel are required to pack aluminum foil fire shelters with them on fires. We affectionately call them "turkey bags" or "shake and bake" after foil bags that are sold to wrap around and bake poultry. The marketed turkey bag is supposed to maintain even heat around the birds in the oven to speed up the cooking process. The fire shelter is designed to have the opposite effect and reflect heat away from a trapped firefighter.

Apparently the Ship Island fire made a run at a helispot containing a lot of firefighting gear. There were two helitack members on the spot: one crossed the creek to get out of the fire path while the other stayed with the gear. There really wasn't that much he could do on the helispot so I'm not sure why he stayed. He told the other firefighter he'd climb into his fire shelter behind the gear if things got that hot. Macho-ism can interfere with good thinking. The gear caught fire, generated a lot of heat, and the fire shelter worked like a turkey bag baking the firefighter.

The investigation team was concerned that firefighters might wear holes in the seams of a fire shelter if they packed it on the line for some time, making it ineffective. I tossed them mine. It had seen a lot of hard miles on the fire line. We could see some daylight through the creases, but everything seemed to be where it should be. Later experiments indicated that a shelter will still reflect heat as designed, even with some daylight showing at the folds as long as wind generated by the fire intensity doesn't rip it apart at the worn seams.

I suggested that there should be a full investigation every time a shelter was used. Someone screwed up if a firefighter finds it necessary to crawl into his shelter. In reality, every trained firefighter has received training adequate to help him determine when it is not safe to stay in a dangerous situation. However, he can get too busy fighting the fire to notice and it is tough to abandon something you worked hard to get in place and run for your life. Overhead personnel must refrain from placing firefighters in situations that involve too much risk but they can't be with each individual crew member at all times. Consequently, it's the individual's fault for not recognizing that things are falling apart in time so he can get the hell out of the way. No one is going to get upset with good judgment. Looking after his safety is a firefighter's first and primary responsibility.

The investigation revealed that this firefighter should have simply followed his partner across the creek to safety. When he dropped into his shelter behind the gear, the burning gear generated heat beyond what the shelter was designed to reflect. He couldn't see what was going on and couldn't move much once he was in the shelter. He hadn't worn his gloves. The shelter got so hot that he couldn't hold it down without burning his hands. Strong winds are present in that much heat which makes holding the shelter in place almost impossible. New standard directions on deploying a fire shelter correctly evolved from the investigation. It is too bad that someone had to die before the need for revised directions surfaced. For me, the most important point the review generated was that an investigation is required every time a shelter is deployed.

We spent another night on the west side of the River and had things pretty well wrapped up there with good assistance from the day shift. The east side of the fire, across the River, was still causing problems. My crews and I, along with needed camp support, established a spike camp just across the Middle Fork from Greyhound Creek by this time to avoid the long helicopter shuttle back to Bruce Meadows. We could get more sleep there then in the main camp with all of the activity during the day and it was easier to fly in supplies daily then to fly us back and forth. We were putting crews at risk every flight since having a helicopter crash is always a possibility.

The day shift was trying to stop the fire at Greyhound Creek on the east side of the river. It was extremely steep and rocky, and they weren't having much luck. Their line needed to be burned out south of the creek but weather conditions made that strategy dangerous during the day.

We volunteered to try it at night. We crossed the river in rubber rafts and were on the line before dark. We improved some line until the humidity started to rise about 10 p.m. We got everyone in place and torched the whole problem stretch. Talk about a major blaze! Fortunately, the humidity was high enough to prevent serious spotting across the canyon behind us. That burnout created quite a spectacle, but it worked like a charm. We lit up that whole part of the world.

I didn't notice that I had gotten so busy lining everything out that I forgot and left my fire shelter in camp until I reached for a canteen that was on the same belt. About that time, a Sector Boss complained that he had gotten so busy that he forgot his lunch. I shared my lunch about midnight, he shared his water, and we discussed what the Safety Officer would think if he found out we were so negligent. Dave Kimpton, the Safety Officer, showed up about then. He thanked us for lighting up the night so he could find his way. He had gotten so busy that he forgot to bring his head lamp. He got stranded in some cliffs shortly after dark, and wasn't able to see to move until we lit the backfire. We promised not to tell anyone about his problems if he would forget ours.

The burnout at Greyhound Creek was a textbook success. We returned to the spike camp about six a.m. I tried to radio the news to the rest of the overhead team at the Bruce Meadows base camp but they were in a strategy meeting with personnel from the Challis National Forest and were not available for the message. We ate rations and crawled into our sleeping bags.

I tucked my fire line radio under my head for a pillow in case anyone called. They did, about the time I got to sleep, and were really tickled to hear the news. Now the only questionable area remaining was in a series of cliffs and broken ground east of the river below Greyhound Creek. The hillside in question had dense Douglas-fir timber pockets scattered through massive rocks and cliffs. The fire had burned up-slope under the timber on the ground, burning some areas and leaving others. The whole area was a maze of spots but most of the fuel on the ground appeared to have burned. The tree crowns had been dried where the fire had burned under them but everything seemed stable. The terrain was too rugged for the day crew, which left us out too. The day shift constructed a good indirect line and burned out what they could on the ridge above the cliffs. The rest of the overhead team and Challis National Forest personnel inspected the line while we were sleeping the next day. They liked what they saw and released our team.

I had been home for a couple of days when the next call came. We were going back to Mortar Creek. Strong winds had swept across the fire area during a very hot afternoon. Somewhere down in those cliffs below Greyhound Creek a pile of logs that remained unburned caught fire in the wind and heat, jumped into the crowns of the scattered fir trees and the world caught fire. The Challis personnel who were there said the fire front was close to half a mile wide when it reached the fire line on the ridge and it went over them like they weren't there.

We could see the huge smoke column long before we reached the Bruce Meadows Base Camp. Someone once told me that the typical convection column on a major forest fire produces as much energy as an atomic bomb blast every five to ten minutes. I suspect this one was generating the equivalent of an atomic blast a minute. If anything, the wind and weather got worse every day. A lot of the rest of the West caught fire in the same wind storm making crews hard to find. I took the crews that were assigned to me and returned to the spike camp on the Middle Fork of the Salmon River below Greyhound Creek. The Challis fire crews who remained on the fire after we left had been forced to place their major effort into trying to catch the conflagration that boiled up through the cliffs below Greyhound. No one had time to patrol the fire line we left west of the river and the fire escaped there as well. I was assigned that part of the fire west of the river, designated as the Dome Division, on day shift.

The fire grew to over eighty thousand acres in just a few days. The terrain, weather and lack of access made our efforts almost negligible. We'd just get one area so it looked secure, turn our primary attention on the next section, and something would take off behind us. Still the politicians were demanding that we put it out. I wish they could have been there to help us, instead of in air conditioned offices in Washington and Boise. It was obvious on the ground that about all we could do was save specific points of interest and spend a lot of taxpayer's money. The fire was burning deeper into the River of No Return Wilderness, and wasn't about to slow up until the weather changed.

We lost ground every day with most of my assigned crews having to retreat to designated safety zones each afternoon. Most of the crews and I had been burned over several times when we received a call from the Boise Interagency Fire Center to see if we could save a major resort owned by Harrah's Inc. out of Reno, NV, and some

Forest Service structures at Indian Creek. There was too much smoke in that vicinity to send in jumpers or helicopters. I checked the map. It would be over a twenty mile hike for us.

I volunteered since we weren't holding much where we were. The smoke lifted before we started so jumpers could get in and we stayed on the Dome Division. The jumpers saved the buildings as the fire swept around them, a politically popular move.

The fire on our division moved higher to the west. We moved the spike camp to a helispot designated H-1 (the same helispot we used when we first reached the fire a couple of weeks earlier) on the ridge between Dome and Mortar Creeks to save a lot of climbing every morning. The fire continued to expand. We'd build line along the flanks in the morning and try to hold it through the hot, windy afternoons. Not much held. We had a brief period around noon when we could safely burn out the fuel remaining between our line and the fire. We routinely lost everything we couldn't get burned out by early afternoon. I spent a lot of time making sure that everyone knew where their safety zone was.

The wind really blew every afternoon and the weather got warmer and dryer. The fire was so large that individual canyons would start to burn out in separate locations. They would throw up individual convection columns that frequently united into a major column that looked like a massive cumulus thunder cloud over the fire. In fact, lightning actually developed from the column under the most extreme behavior.

That type of activity created very unpredictable burning conditions on the ground. I tried to stick with the crews working in what looked to me like the most dangerous areas. The fire usually flared up and chased us into our safety zones every afternoon. We spent almost every afternoon perched on rock piles or talus slopes, or in a pine grass/elk sedge vegetation type that wasn't dry enough to burn yet while the fire swept over and around us, negating everything we tried.

I was working on the line with some crews one afternoon. The wind shifted, and the convection column shifted with it across the steep slope in the direction of our spike camp. Small spot fires produced by burning material falling from the convection column flared up all around us. I radioed everyone in that area to have them try to catch the spots, and reminded them to re-assess their safety zones, then headed for camp. We had a lot of camp crews, timekeepers and other administrative personnel in the camp. Most of them had never had a fire run around and over them.

Things didn't look good in camp. I lined everyone up and explained what was going on. We had one especially vocal time keeper there who was sure he deserved hazard duty pay just because he had to ride a helicopter to get to camp. Workers on the fire line qualify for such pay up to the time when the fire is declared controlled. Camp personnel don't qualify. I advised him he might be able to justify the twenty-five percent increase in pay before this day ended.

A Bureau of Land Management hot shot crew from the Boise Interagency Fire Center returned to camp about then since it represented their safety zone. The spot fires they had been trying to stop were consolidating and there was nothing they could do to stop the spread. I radioed the other crews involved to return to camp or go to other safety zones, whichever was closest. I had everyone gather up all of their gear and stack it in the center of camp. The hot shot crew briefed the camp crew again on use of fire shelters. I located the best area of elk sedge and pine grass and sent the camp personnel there.

Spot fires were popping up all around us, with a major crown fire coming upslope right at camp. I helped the hot shot crew set everything on fire around the camp with back fire fusees. We hoped to break the intensity of the fire coming at us by getting a good black area around camp.

The Line Boss flew in just ahead of the fire to help size things up. He agreed that we were doing what we could to keep everyone safe, and returned to his helicopter. He suddenly came back. He had room for one person on the ship, and asked if I wanted to go. The most important problems I had at that time were right where I was. I declined. I asked if anyone else felt the urgent need to fly out. No one volunteered. I almost sent the timekeeper who was so worried about his hazard pay. Then the helicopter was gone and we were there for the duration.

The fire hit us head on. I walked the perimeter assessing the need to go into shelters while the hot shot crew used back pack pumps filled with water that had been flown in for drinking to keep the gear from burning. The burnout from camp worked just as we had hoped. No one had to go into their shelters although the fire went right over us. We all had minor burns from sparks that got down the necks in our shirts, or between our gloves and the sleeves of our fire resistant shirts. We inhaled a lot of smoke. The heat along the perimeter got hot enough to melt the plastic name tag on my hard hat. We had over seventy people in camp when it burned, but we managed to keep everyone calm and they all made it.

The only gear we lost belonged to the hazard duty timekeeper. He wasn't the sociable type, so he camped as far from everyone as he could and still feel safe from the critters he knew were out there waiting to eat him. He got so excited about what was going on that he neglected to tell us where his gear was. He sent in a major compensation claim that I recommended not be accepted. We could have saved everything if he had done his part.

We had one medical problem relating to the burnover. A Job Corps member on the camp crew had asthma but his pride and commitment to his crew wouldn't let him leave the others-the fellowship is that strong. He would have gone out on the empty slot on the helicopter if we had known. The smoke and excitement put him down. We had to evacuate him by helicopter as soon as the smoke cleared enough to get one in.

We'd hit the line each morning, build line along the flanks, and try to hold during the heat of the day. A night shift was sent in to try to reinforce and hold what we accomplished each day. We made some progress, but it was slow going in rugged terrain and impossible weather. The fire was making far more progress than we were, out distancing us with runs each day. We continued to lose line. It was almost impossible to pick up new crews as the old ones wore down.

Somewhere in all that smoke I received word that I had a major fire going back on my ranger district in Nevada. The Shanty Town Fire started on private land near Ruby Marshes and burned over nine thousand acres during the first few hours. I couldn't get more details, and what

I heard was a few days old by the time it reached me. All I could do was lead folks back on the line each morning and hope we'd eventually catch the Mortar Creek Fire before Christmas.

We got a minor break in the weather one day when the wind died and the temperature cooled a couple of degrees. We weren't getting the plans from base camp until after our crews went on the line, and were pretty much on our own. We pushed hard to take advantage of the weather. I don't know how many miles of fire line it took to tie in the north side of the fire and cut the lead, but we made it. We ran out of time and it was still too hot to burn out the line to make it safe. We had to leave that for the night crew. They had been having problems too, but I was sure they could do it. The night Division Boss agreed to give it a try. I met him on the line the next morning. He didn't blink an eye when he told me they decided it really wasn't safe, so they hadn't burned a thing. The fire had crossed our line in one short section, and that apparently spooked them. They did get that re-lined, but across a slope where we wouldn't be able to hold it during daytime burning conditions. I really can't find fault with them for not completing the burnout. They were getting as worn down as we were and got even less sleep.

That left us facing an impossible task with impossible burning conditions in the heat of the day. We had to try. I got everyone going as soon as I could. We relocated the line built during the night, with only one delay when that crew ran into a yellow jacket nest.

We had a "Mark-3" pump "long-lined" in to us by helicopter in the bottom of Dome Creek. The pump was attached to seventy feet of rope (a long line) under the helicopter, and lowered to a point near where we wanted it. We torched everything for the burnout as soon as we got everyone in place. Surprisingly, everything seemed to be going great. The main fire was really heating up with good convection column development that was sucking our backfire right into it. I actually had time to think we might pull it off.

Then the wind picked up. The huge convection column and all of the fire brands within it bent right across our line like a bad dream.

I called all of the Crew and Sector Bosses on the radio even before the resulting spot fires started picking up and advised them to head for their safety zones. Things were going to hell fast when I got to the pump which had been set up in the sector in the most danger. The person manning the pump asked what we should do with the equipment. It was too heavy to carry with us in the available time. I helped him throw it into the creek and we all took off running. The safety zone on this end of the line was a patch of talus rock below a cliff. Everyone had time to make it as the fire closed in behind us.

I was just feeling proud of everyone for reacting as they had been trained when a Crew Boss from the Olympic National Forest came to report that he couldn't account for a male and a female on his crew. He wanted to go back into the fire to look for them. He was from the Cascade Mountains, an entirely different area with an entirely different background. There was no way that I could let him go back into the fire.

We didn't have much time to find them. I knew the fire, the fuel and the terrain, and I was in charge. I hadn't lost a firefighter on other fires, and wasn't about to lose these.

I got the first names for the lost people, told the Crew Boss to stay with the others, asked everyone with radios to try to contact their counterparts in other safety zones to see if they could locate the missing couple, and went back into the fire.

A person can't imagine the noise, smoke and confusion created by a big inferno when it makes a major run unless they've been there. I was running right back in front of it all. I shouted the lost firefighters' names, jumped logs, and moved as fast as I could downhill into the area where they had been working as spot fires started to run together. Trees were starting to crown all around me.

Visibility was shutting down as the smoke and fire intensity increased. I doubt that my voice carried very far. The fire was making too much noise. I held my radio near my ear as I ran, hoping against hope that the lost people would show up somewhere and someone would notify me on the radio. Crew Bosses and others are assigned different radio frequencies so everyone has a chance to use the radio when necessary, so I wasn't hearing everything.

I pushed my way into the fire much farther than I knew was prudent. I couldn't get the request that Lois had made several years before out of my mind. I was taking far too many risks to try to rescue a couple of irresponsible firefighters who had chosen not to follow orders. Yet I somehow couldn't leave two people with little experience in the middle of a fire. I picked out the trunk of a large spruce tree through the smoke about fifty feet ahead. It was just above where the pump had been, and where the two had last been seen. I decided to turn back at that tree, even if I couldn't find them.

Then I heard my name on the radio, and responded. The

two people had been found. They failed to follow orders to maintain contact with their crew and ran the wrong way. They showed up in another safety zone by following another crew. I doubt that they were smart enough to know where that left the rest of us. At least they weren't the crispy critters they would have been if they hadn't gotten to safety.

I turned back as fast as I could. It was an uphill run now, and the fire was really building up. All of the logs, brush and rocks encountered on the downhill run now had to be negotiated through the building conflagration. I was forty years old. No one, especially someone that old, should have to run that hard uphill in that terrain, but there was quite an incentive. I am sure that I could not have made it if I had continued on to that spruce I had chosen as my final target. It was just too far into the fire. A pile of rocks never looked as good as the talus slope did when I finally reached it. I spent quite a bit of time on my knees retching and trying to vomit although I really didn't have anything on my stomach. Catching my breath was tough because of all of the smoke, the biggest problem. There was almost as much of it on those rocks as there had been in the timber. The lukewarm water in my canteen had never tasted better after I quit trying to heave. I even took time to thank God for letting us all make it to safety.

Smoke from a fire like that one does weird things to the light. The sun just disappears. When it does show through you can look directly at it. It's a surprisingly big circle. The air turns an amazing luminescent greenish-orange. We fastened our top shirt button in an effort to keep sparks from getting down our neck. I teamed up with a Sector Boss in an effort to locate the sparks that lit on our fire resistant clothing and put them out. Nomex clothing won't burn, but that doesn't keep a spark from burning through.

There wasn't much to do. We just had to sit it out. I called around to see how other crews were doing. Significant interference can develop with radio transmission in the middle of a fire. I assume it is created by electrical activity in the huge convection column that surrounded us. I managed to reach everyone although it took several attempts.

One sector had pulled into a black area that had crowned out previously. They were safe, but concerned about all the smoke that was creating breathing problems. I assured them they weren't alone. The Sector Boss suggested that they would feel more secure in their fire shelters. I advised against it. The area wasn't going to burn again and crawling into a shelter doesn't produce more oxygen. It does create claustrophobic conditions. A firefighter can't see others around him, so he feels all alone and insecure.

The safety zone for another sector was an area of the elk sedge/pine grass. They were safe, except for all the smoke. This was the group with the two stupid firefighters who abandoned their crew and ran the wrong way. I still have trouble forgiving them for coming so close to killing me.

I had another sector that was working the southeast flank well below the run, and were still working the line. Things were pretty hot for them, but they were holding.

The main fire eventually worked its way past us but things were still very smoky. I advised the crews to sit tight for another hour before trying to pick up the flanks where we had lost everything. They were a pretty dejected bunch. Maintaining moral in tired crews who have been burned over day after day is tough under the best circumstances. It was obvious to everyone that we weren't making any progress on controlling

Mortar Creek and that it was growing larger every day in spite of our best efforts. It wouldn't be easy to get them to make much progress again that afternoon.

I left them in their safety zones and headed down through the smoke and burned out landscape to see if I could do anything to help hold the southeast flank. They were barely holding and not making much forward progress. A tree had suddenly crowned beside one young lady, scorching her hair where it hung outside of her hard hat. She didn't think it was funny when I suggested that was supposed to help eliminate split ends.

I ended up with the rest of the crews in camp before dark. The big news was that my overhead team was being relieved. I don't remember how long we had been on the fire line, but it was at least three weeks. We were more than wore down. I had developed a severe cough with lung congestion from all of the smoke, and it was tearing me apart. The run up the mountain through the fire while looking for the irresponsible fire fighters hadn't helped. The rest of the overhead team, and several crews, weren't in much better shape.

A fire had never whipped me before, but Mortar Creek did. The fellowship of the other firefighters had helped support us all for the long period that we had been on the line. We were all brothers and sisters with a common goal, one we could not achieve. I wasn't a happy firefighter as I gathered my gear and waited for the helicopter the next morning. I cried on the flight out as I looked down at the miles of fire line we had lost. The brown scars identifying our lost fire line were all too obvious through the blackened landscape below the helicopter. We had tried so hard, but the Mortar Creek Fire was not to be tamed by man.

CHAPTER 29

Nevada Fires

It was a long trip back to Nevada to see how my district fared in my absence. Other teams and crews didn't get Mortar Creek under control until snow blanketed those mountains late that fall. It burned thousands of acres of wilderness and cost the taxpayers several million dollars to try to control. The two people who started it were caught but they obviously lacked the assets necessary to cover control costs.

One of the first things I had to do back on the Ruby Mountain Ranger District was to review damages done by the Shanty Town Fire and see what we could do to prevent erosion and re-vegetate the burned area with desirable plants. Cheat grass, an undesirable plant introduced from Asia, was present in the vicinity. It will dominate burned areas if given the opportunity. A critical deer winter range had burned which caused special concerns. That Nevada country burns well under the right conditions. There isn't much vegetation on a lot of it. However, take a couple of months without rain, add one hundred degree plus temperatures, drop the relative humidity to five percent, kick the wind up to fifty miles per hour, and it doesn't take much fuel to create a major conflagration.

Those were the conditions firefighters faced on the Shanty Town Fire. Mostly they just had to keep out of the way until something changed. One Nevada Department of Forestry employee got brave and raced out in front of the fire with his new fire truck. I'm not sure what he hoped to accomplish against a fast moving fire all by himself with only a couple of hundred gallons of water. I'd give him A for effort and F for smarts. He high centered the truck on a rock then jumped out and ran out of the way. The truck burned well. He was a very lucky man. The wind died down and temperatures dropped when the sun set and crews were able to catch it that first night at 9,400 acres.

I got a call the first Sunday home: we had a major fire on the Bureau of Land Management (BLM) public land between Harrison Pass and Overland Pass on the west side of the Ruby Mountains. It was burning fast and headed for the forest. I had no trouble seeing the convection column from Interstate 80 en route to catch a plane in Elko to fly the fire.

The fire had reached the forest by the time I got over it. It was burning fast through Pinyon-juniper and sagebrush/grass fuels. I could see several cattle moving into a fence corner where they would be trapped but was able to contact a Nevada Division of Forestry truck on the radio and requested that the crew cut the fence so the cows could get out of the way. About all we could do was build line on the flanks of the fire until things cooled down.

We landed the Cessna on a dirt road where Walt Grows, my assistant from Lamoille, picked me up and we drove to where the fire started. It was a clear-cut case of arson. We found at least three starts, plus one that didn't take. The fire bug used paper matches. I called for an investigation before turning my attention to the fire.

Lots of "sagebrush rebels" were cropping up in Nevada. Most of them were local ranchers and miners who didn't approve of federal restrictions on what they could do on public lands. They believed that the government was out to take away what they thought were their constitutional rights to do what they wanted on public lands. Some were pretty extreme. They demanded that all public lands be turned over to private or state ownership. I provided the fire investigator with a list of about half a dozen names that should have included the arsonist but he wasn't able to prove who was responsible.

This fire started along an old stage route between the historic Overland Trail and mines farther to the north as far as Idaho City, Idaho. A major stage stop along the route, Cass House, was next to the blaze so we named it the Cass House Fire. The media picked it up as the Cat House Fire which apparently caused quite a ruckus in some circles. Cass House probably offered those services too, so the misnomer may have been appropriate.

I ordered a couple of crews but kept the staffing light. The fire was burning towards solid rock at timberline on the mountain crest, and that makes a pretty good fire break. Resource values at risk just didn't justify major staffing. The rancher who ran cows in the area came up and bawled me out for cutting the fence. Now he would have to fix it. I told him I thought that cutting the fence beat cooking his cows but got the impression that he thought cooking was better. He could have sued the government for neglect and would have made a lot more money that way then from selling those scrawny critters at auction.He was also upset since I wasn't spending thousands of taxpayer's dollars to keep a couple of week's worth of grass for his cows from burning. In actual fact, the fire would create much better future forage for his cows.

It took a couple of days to get the fire controlled. I intended to spend just one more night then leave it for the crews to mop up. Another call came in before dark. We had another major fire taking off in Ruby Valley on the east side of the mountain range. I jumped in my truck to see what was happening there.

The fire started near the Rock House, another historic structure that now served as a local general store/bed and breakfast. The person who started this fire, a geology student staying at the Rock House, was quick to confess. She had been standing in some tinder dry cheat grass, trying to chip quartz crystals out of a rock with her rock hammer. Apparently a spark from the hammer ignited the grass. It seemed like an unlikely story, but discussions indicated that she didn't smoke and hadn't done anything else that could have started the fire. She had no reason to lie. She was just chipping away, felt something warm, looked down at her feet, and everything was on fire. Slim Saxton, owner of the Rock House, rushed out to help but it was well out of control before they could do anything. So I ordered a couple of more crews, and found a new home at the Rock House Fire. We flanked the east side of the fire to keep it away from higher values in the valley, and ran it into the solid rock timber line to the west.

Now I had three major fires on the district. I needed to get with experts in hydrology, vegetation, soils, wildlife and other specialties to see what was needed to rehabilitate the burned areas. I served as Incident Commander (fire boss) until everything looked secure on the Rock House Fire a couple of nights later, then headed for home. I had a meeting set up with necessary resource specialists in Elko for the next morning. Once we identified rehabilitation needs we could seek funding.

That part of Nevada was full of black tailed jack rabbits that year. I couldn't miss running over some of them with the pickup truck as I drove up the road in the dark. I reached U.S. Highway 93 about twenty miles south of Wells about two in the morning. That's a lonely stretch of highway, especially in the middle of the night. At least it was paved.

I slowed down to about fifty-five miles per hour as I approached some irrigated land and ranch houses. Some ranchers run cattle on the extremely dry desert BLM range along the unfenced highway in that area. It's very marginal range with little for the animals to eat. The cows are attracted to the lush irrigated fields, and tend to concentrate along the fences. I was concerned that one of the cows might be on the road.

I had seen headlights in the rear view mirror for a long ways which turned out to be a big truck hauling heavy equipment. He must have been going close to a hundred miles per hour, heading from Ely towards Idaho. He came screaming past as I slowed because of my concern for cattle close to the irrigated land. His lights cut down my vision as he passed. I remember thinking, "There go all of the cows from here to Idaho."

He was still in the other lane beside my truck when I noticed something in my lane. At first I thought it was just another rabbit. It kept moving and flickering. It took an instant to identify a young black Angus bull as he turned his head. He was walking right down the middle of my lane about thirty feet ahead of my bumper. The flashes I saw were caused by my headlights reflecting off of the white alkali dust on the bottom of his hooves.

I stood on the brakes-too late. I hit the back side of the bull just to the left of the truck's center. The bumper jammed against the left front tire, the hood flew up, the battery exploded and the lights went out. I knew that I was going off on the left side of the highway, since that tire was locked against the bumper and I couldn't do anything about it. At least the big rig had just pulled clear.

I remember hoping that I wouldn't roll, and then was amazed at how well the shoulder strap on the seat belt in that Dodge pickup had tightened up. I felt pretty secure. I skidded to a very dusty stop in the greasewood flat. The pickup didn't roll.

The big rig kept going. He either didn't want to get involved, or figured it would take him a couple of miles to stop at that speed. It's a good thing I wasn't hurt. The radio went out when the battery broke. I doubt that I could have reached anyone at that time of night anyway.

I could see another set of lights coming about ten miles back down the road. I checked the highway to make sure I wouldn't cause another wreck. I had knocked the bull off of the highway with the truck. He was coyote bait. I kicked chrome and glass off of the highway and waited for the other car. It turned out to be a little Ford Mustang. I'll bet it had at least a dozen Mexicans in it.

The local white welfare leeches were too good to wash dishes or make beds in the truck stops and casinos, or to change sprinkler pipes on ranches. Welfare paid better, they got to stay home and make babies, and they didn't have to pay taxes. That left the casino operators and ranchers depending on green card or illegal aliens to do the job. A lot of illegal aliens were transported through that country at night. They were much better workers, and were much less expensive than the welfare people would have been if the business owners could have gotten them to accept a job offer. I was in a federal uniform and felt lucky they hadn't run me down.

There was no way that we could pile one more person in that small car. I'm sure they would have added another person already if that had been possible. Between the Spanish I knew and the English they knew, we discussed what to do. I understood there was another vehicle behind them that had more room. I thanked them and they took off. I could see the other headlights before they left. This time it was a six-passenger Dodge pickup. It was equally full of bodies, but the only thing in the back was a couple of huge dogs. It took a bit of doing, but I made adequate friends with the dogs to join them in the cold back of the truck.

Jim and Anna Whited ran the ranch just up the road. They were good friends, which helped when I woke them up at about three a.m. Jim met me at the door with a gun, but decided not to shoot when he realized who it was. I made necessary arrangements on their phone and got home about daylight.

Tom Ross was a young Presbyterian minister who drove a three hundred mile circuit to serve three churches each

Sunday. When Tom heard the story he commented that he frequently saw the large smoke columns coming off the mountains and desert and knew they were wildfires. He just never thought about all of the people out there day and night trying to control them. He promised to pray for us. I encouraged him since we needed all of the help we could get.

We had a couple of smaller fires before snow fell. I had more than reached even my tolerance for fire season by then. I was very happy to see the snow.

CHAPTER 30

Incident Commander in Montana

In spite of all the fires, I had accomplished the major tasks the Forest Service had sent me to do in Nevada. The personnel problems had been solved, although it took some doing. It isn't that easy to fire or move ineffective employees, or to get them to resign, in the federal government. I did all three. Now we had a very effective crew on board.

Most of the good ranchers had been more than anxious to improve their range, so we had implemented a lot of new grazing plans. There were still some holdouts, but we had included most of them in improved programs as well.

I started putting in for new jobs. The staff officer in charge of range, wildlife, fisheries, minerals, soils and watershed on Montana's Beaverhead National Forest was open. Somehow I managed to get it!

We loaded up the family and furniture and headed for

Dillon, Montana the last week in July, 1980. At least nothing froze that trip.

We had a few fires in the Beaverhead area after we arrived in 1980. I was too busy learning the new job to make much effort to go on one. I waited until 1981 to start making sure that everyone knew my qualifications and that I was interested in fire.

Major state and federal agencies charged with handling serious incidents such as forest fires were undergoing a consolidation in organization and terminology during this period. We integrated into the Incident Command System (ICS) which had worked well in California for some time. When a major "incident" occurs, be it fire, flood, earthquake or whatever, someone has to organize things. That's where ICS comes in. The system recognizes that no one agency has a monopoly on expertise to handle emergencies and, by pooling resources, agencies can have more expertise and equipment available while reducing cost, training and duplication of effort.

Montana had three interagency Type Two Overhead Teams on the "east side" (east of the continental divide). I was assigned as Incident Commander for one of these teams in 1982.

A Type Two Team includes an Incident Commander (old Fire Boss), two Operations Section Chiefs (Line Bosses), a Planning Section Chief, a Finance Section Chief, a Logistics Section Chief, and a Safety Officer. They all have their own internal organization to handle specific needs. Key members on my team worked for the U.S. Forest Service, Bureau of Land Management, Bureau of Indian Affairs, and National Park Service. State and other agencies were eventually added. We handled incidents of limited complexity since the standard internal organization wasn't as extensive as that in Type One Teams.

I had a great team and could hardly wait to try it on the ground. Unfortunately, 1982 was a wet year with few fires. We finally got one call, for the Greek Creek Fire on the Gallatin National Forest.

The fire started next to the main highway in Gallatin Canyon, between Bozeman, Montana and Yellowstone National Park. It was burning in a concentration of dense dead lodgepole pine that resulted from an insect attack that originated in Yellowstone Park in the late 1970's. One of the toughest safety problems involved managing traffic so someone didn't get killed in that narrow canyon full of speeding vehicles. Everyone wanted to stop in the middle of the highway and gawk at our fire. The fire caused almost as many traffic jams and potential accidents as a bear sighting does in the park. During the summer, that's one of the most overloaded highways in the nation.

We used a helicopter to place some portable water tanks on a ridge above the fire, filled them with buckets slung under helicopters, set up a water relay system, and caught the fire in short order. The biggest problem involved pressure buildup in hoses working downhill from the tanks. We didn't need a pump. We simply started a siphon at the tanks and built up too much pressure with gravity before we reached the bottom of the fire.

I thought we had pulled this fire mission off without an injury, in spite of heavy traffic, steep, rocky slopes, and lots of snags. I was helping control traffic and load firefighters on buses when the last crews came off the mountain. The very last man on the very last crew was packing a chain saw. He tripped as he stepped onto the cut bank for the highway and rolled onto the road. Gas from the saw spilled into his eyes. Somewhere in that fall he managed to jam a stick in his ear. There was an emergency medical technician on the crew. We immediately washed the gas out of the victim's eyes with water from canteens. The stick was wedged tight in his ear. We immobilized the stick as well as we could and sent him via helicopter to the hospital in Bozeman. The doctors indicated he would have lost hearing in that ear if we had tried to remove the stick in the field.

Greek Creek was the only call any of the east side overhead teams had that year. Although I had lots to do in my staff officer position, I was looking forward to tackling more fires with that team.

CHAPTER 31

BIA Fires

Conditions changed in 1983. Spring started off quite wet in Southwestern Montana but the eastern part of the state was extremely dry. I kept a radio with me but missed the first fire call when I was in an area lacking radio reception during a tour evaluating livestock use in the Gravelly Mountains. I didn't know that the radio repeater covering that end of the Forest wasn't working. They had to fill in with a substitute Incident Commander and the team went to the Northern Cheyenne Reservation near Lame Deer without me.

I made the second call. Six fires were burning on the Crow Reservation and the agency wanted us to take them as a complex. Three fair sized fires were still moving when we got to Crow Agency, Montana. The fire with the most potential was burning on Reno Ridge, where Major Reno made his stand after trying to support Custer in 1876. The fire with the most publicity burned the Little Big Horn Battle Field National Monument.

This was my first experience as Incident Commander on a reservation fire. One of the Operations Sections Chiefs worked for the Bureau of Indian Affairs (BIA) on the Rocky Boy Reservation to the north. I kept him handy. Trying to work through the politics between the Federal BIA and tribal councils can be interesting. The BIA has to deal with standard national direction with significant concern towards meeting individual tribal needs, while tribal councils are bound by demands from reservation residents and local political pressures. Politics make both jobs difficult, and unfortunately, objectives can differ on specific issues.

Most of us have a tough time understanding the strong family and cultural ties that direct life on reservations. Unfortunately, some of the related problems tend to promote high levels of unemployment and related social problems. Residents receive welfare payments and related "assistance", giving them little incentive to work. Lots of people with excellent potential are essentially trapped in those environments. I see the same problems, minus the strong family ties, occurring in major city ghettos. I wish we were smart enough to develop a solution to these social problems.

Some members of the tribal council suggested that we use a vacant motel near the Monument for the base fire camp. I inspected it with the Finance and Operation Section Chiefs. It was a fairly new structure, constructed with federal funds to increase reservation employment opportunities. It represented what happens with a lot of those welfare projects if the government doesn't gain sufficient local support and commitment in advance. Although still habitable, it was essentially in ruin, the victim of extreme vandalism and neglect. I don't consider myself an overly suspicious person, but the possibility existed that, if we took the offer, a claim could be submitted that the fire crews did the damage. We declined the offer.

The agency was most hospitable and made several new houses available that had been built with federal funds but were currently unoccupied because the contractor that made them used substandard material that failed to meet construction standards. They made a good camp. Essentially everyone was busy at Crow Fair, a major tribal celebration, so we had things pretty much to ourselves.

The largest smoke on the Little Big Horn Battlefield Fire was coming from a fire truck when we arrived. A crew had driven the new truck into the path of the blaze and started stringing hose. No one stayed with the truck. The fire flared up, they felt threatened and abandoned the truck.

Reno Ridge and a troublesome fire to the north were still going strong. That's great bulldozer country. Some of those Native American cat skinners are among the best I've seen. They are utterly fearless. I'd just decide we couldn't safely take a cat somewhere and one of them would, building fire line all the way. It took several days to contain the fires in the high winds we faced. Like the Nevada fires, there isn't much mop up once we stopped them. That country sure was dry. I figured we'd be back shortly.

CHAPTER 32

Crazy Head

Our next call was to the Northern Cheyenne Reservation out of Lame Deer, Montana. We were assigned the Crazy Head Fire. There were a lot of other fires going in Montana when we arrived. Dave Filius, the Incident Commander on one of the other East Side Type Two Teams, and his crew were assigned the Horse Head Fire just south of us. Both fires were moving fast. We set up our base camp at Crazy Head Springs.

Montana Indian Fire Crews (MIF) are composed primarily from reservation residents.Unemployment runs high on reservations so there is significant competition to get on a fire crew. They put together some excellent crews, especially with the first several to be assembled. If there are lots of fires going around the west and most of these experienced crews are out, some MIF crews are more like the old pick up crews encountered in the late '50's. Politics were such that we normally had only MIF crews on reservation fires, with the possibility of slipping in a type one hot shot crew on some of the more serious blazes. We had good MIF crews on Crazy Head.

The weather wasn't cooperating and both Crazy Head and Horse Head Fires continued to grow. We faced a major logistic block by having to go through the agency dispatch center in Lame Deer. They did their best but weren't equipped to handle that work load. We had a lot of competition for supplies because of all of the fires burning in Montana and other western states.

After discussing supply needs with several BIA personnel and tribal members, we realized that different units within the agency worked quite independently. The agency fire organization didn't know all of the resources other units had. Dave and his team were having the same problems we were in getting supplies for our separate fires. We started checking farther and determined that some of the supplies we needed were stashed away within other resource management units at BIA headquarters in Lame Deer. These units hadn't volunteered their supplies due to concerns about the logistics of replacing them after the fires were controlled.

Probably the best example involved the lack of anything to burn out the lines as we constructed them. Dave and I discovered that the fire unit in Lame Deer was out of backfire fusees, but the timber unit had some stored in their area for burning logging slash in the fall. We made a midnight run with someone with a key and found several cases of fusees. We also located several drip torches that were even better. A drip torch uses a mixture of gasoline and diesel. The person using the torch lights a wick on a spout, tips the torch up on end by a handle and starts to walk through the area to be burned. The flammable mixture "drips" through the wick and burning beads of liquid drops to the ground and starts the desired backfire. Dave and I split this find between our crews and burned out a lot of line.

We had trouble holding our lines in the strong winds that blew every afternoon and sometimes late into the night. The fires grew big enough to show up on weather satellites which concerned fire personnel at the interagency coordination center in Missoula. It is tough to get much information passed up the line on reservation fires. The fire center in Missoula was concerned that the fires had grown so large that we couldn't handle them with Type Two Overhead Teams.

Crazy Head and Horse Head had almost burned together by then. Someone called from Missoula to ask what Dave and I would do if the fires burned together. We both responded that would just give us one less flank to worry about; we would handle the combined fires as two zones. That seemed to be the right answer since we weren't replaced with Type One Teams.

The tribal council requested that we keep the fire crews out of specific areas. One was a tree where a tribal member had died the previous winter. Disoriented driving back to Lame Deer in a blizzard, he got on the wrong road, got stuck and tried to walk out. He froze to death under the tree. The family tied various "religious bundles" to the tree which was easy to see and avoid. However, it was between Crazy Head and Horse Head. The area almost burned on several occasions and the family eventually moved the material.

The second area was the tribal Sun Dance Area, a sacred ceremonial site. The fire eventually ran over it. We kept the crews away as requested. I flew over the area after the fire had passed. The blaze swept around the site and never touched a thing. It may have been because the area had been trampled by the use it received during various rituals. The Cheyenne firefighters assured me it was spared because it was holy.

The third area, the sacred buffalo herd pasture, was just west of the fire camp. It is basically a large semi-timbered pasture where buffalo transplanted from Yellowstone Park are kept. We kept the fire out of there, but it was close.

Crazy Head Springs, site of our base camp, is a popular tribal recreation area. It was closed to non-fire use while we had the fire camp there. We routinely had problems with some very intoxicated folks who showed up from time to time. They would try to lure crew members down to join them in drinking bouts, or start fights. I was concerned about security and the potential for people to get hurt. I asked the tribal police if they could send someone out to represent local authority and enforce the area closure. They sent out a young lady in uniform. She was a small, good looking young lady who seemed competent. It wasn't long before three huge, very drunk men drove up to the spring. They got out of their car and started shouting obscenities at the camp.

The little police lady sauntered down and politely asked them to leave. The biggest guy didn't take kindly to the request. He let out some sounds that would have done credit to a grizzly bear and started for the police lady. I grabbed a shovel and started down to see what I could salvage. A couple of other team members started to accompany me. We didn't need to bother. The young lady landed an expert blow to the big guy's solar plexus with her night stick. He folded up around the stick, trying to catch his breath. She reared back and brought the stick down on the back of his head with a crack we heard quite clearly a hundred yards away. It sounded like she had hit a pumpkin with a baseball bat. The blow knocked him down. He just lay there twitching like a fish out of water. She turned to the other two men and calmly stated, "Get that son of a bitch in the car and get the hell out of here." They understood everything she said. They dumped what was left of the big guy in the back seat, jumped in, and roared away in a cloud of dust. I hope she didn't kill him.

The fire kept taking off where it wasn't supposed to be, especially when we thought we had a particular section under control. We started paying more attention, especially when our base camp almost burned up. Dave and his crew were having similar problems on Horse Head. Someone was sneaking around starting fires outside our lines.

A fire took off early one afternoon just north of camp. The wind had shifted enough that camp was in immediate danger. I called the day Operations Section Chief to see if he had anyone available to help save camp. He and his crews were trying to keep the fire from closing Highway 212 a couple of miles to the east. We had a bulldozer in camp that had just been repaired. I set the operator to building line while camp personnel prepared for evacuation if necessary. I called for a retardant drop, rousted the night shift crews and we tackled the new threat. The sacred buffalo pasture was also in danger. By this time the new blaze was throwing up a very ominous convection column. We initiated a burn out program immediately behind the bulldozer as it constructed line. The convection column sucked our backfire directly into the cloud. Everything burned out in a textbook sequence.

The retardant plane showed up just in the nick of time. His drop was exactly what we needed to hold the line. The only problem was that the drop also coated the haystack used to feed the buffalo during the winter with bright red retardant.

I pointed out the red haystack to the agency superintendent when he reviewed our progress that evening. He took one look at the red pile of hay and stated, "Good, now maybe we can catch the sonuvabitch who keeps stealing our hay."

The next unforeseen blaze made a major run towards Highway 212 after we controlled the fire problem near camp. We closed the highway to non-fire traffic and I took the night crew to that hot spot. We used the highway as the anchor for our burnout. I have never seen a backfire work better.

We stopped the fire at the highway. However, telephone service to Lame Deer was disrupted as the backfire burned up several plastic boxes relating to the telephone line. We had to radio our orders directly to the fire center in Billings which actually helped reduce some of the relay problems we had in ordering supplies.

One of the crews jumped three men out of the brush along the fire line while we were trying to hold the fire at the highway. The men jumped in a beat up old red Ford pickup truck and roared away. The crew didn't think that much about it until they found a drip torch filled with straight gasoline where the men had been hiding. They radioed in the information, and I called the tribal authorities.

The tribal police located the truck and a high speed chase ensued. The men were very drunk and ran the truck into a tree on a turn. Two of them took off on foot and were still running last I heard. The driver was hurt badly enough to get caught. It turned out that the three were having a big argument with tribal leaders, and were starting the fires outside our lines to get even. It's amazing they didn't blow themselves up using straight gasoline in the drip torch.

We kept working and gradually made headway. I was flying the line one afternoon with the Operation Section Chief while the fire was moving through some pretty rough country. We called in a retardant drop for a problem area. The plane was just getting ready to drop when we got word that one of the bulldozers was on fire. It didn't take long to locate it. I think the operator's last name was Elk Shoulder Blade. He was one hell of a cat skinner for a crazy man. He was building line down a steep slope that looked like a suicide run to me. He got too close to a hot spot in the fire and old oil, grease, hydraulic fluid and other petroleum products on the engine caught fire from the heat. He was beating on the burning tractor with his coat when we located him. We directed the retardant drop onto the tractor. The retardant turned the cat from a rusty yellow to smoky red. It also put the fire out. Elk Shoulder Blade jumped back into the driver's seat and tied in that section of line before dark.

Dave was making progress on Horse Head as well. We got things tied off at about the same time, thanks to having stopped the folks who kept setting the extra fires. It had been a tough run, but we made it.

We were having some problems getting the fires moped up. The weather remained hot and the crews were tired. We brought in infra-red heat detection equipment mounted on helicopters. The equipment operators flew the line, located hot spots and dropped cans of frozen juice with bright engineer flagging tied to them on the spots. It was amazing how fast the crews could get to those spots (and cans of frozen juice).

CHAPTER 33

Lame Deer

There were lots of fires after Crazy Head, which kept the teams running that summer. We weren't home long before we were called back to the Northern Cheyenne Reservation. This fire was burning right up against Lame Deer on the west side of town.

The fire burned hot with an unusual wind from the east. I was sizing up a run it was making up a brushy draw while waiting for crews to arrive when a pickup full of men roared up with rifles sticking out everywhere.

"You see a deer?" one of them shouted.

I told them I hadn't.

"Somebody said there was a deer up here. We figured the fire would chase it out!" he shouted, and they roared away.

Big game animals get hunted yearlong on most eastern Montana reservations, with no management to build numbers. I never saw a big game animal in all the time I spent on "east side" reservations. I heard lots of complaints because there was nothing to hunt and tried to point out that they had the habitat during discussion with tribal leaders. All they needed was hunting regulations to manage wildlife populations and they would have lots of animals to hunt. No one made the connection.

It was late September by now. The wind continued out of the east. Weather forecasts emphasized a huge low pressure system to the west (accounting for the east wind), and kept promising storm. The weather continued hot and windy, and the nights were cold and windy. We finally caught the fire.

The dispatch crew in Billings advised us that Billings was weathered in. They sent a plane to get us home if the weather permitted. We rolled bales of hay off the landing field at Lame Deer so the plane could land. The sky kept getting blacker and blacker. We were happy to see the plane when it finally arrived. It was just starting to rain when we took off. The pilot reported extreme turbulence, with a serious storm system directly west. We headed north, dropping overhead team members off at Lewistown, Havre, Browning, Lincoln, Butte, and finally Dillon where I got off. The plane had one more stop at Livingston to drop off a Park Service Operations Section Chief. I've flown a lot of fire patrols behind thunderstorms, but that was one of the roughest rides that I remember. I never get motion sick, but sure felt like a tournament ping pong ball when I stepped out of that plane.

It was snowing in Dillon when I got there. And the storms continued. It had been a busy season.

Lois reported that she was getting worried. She was

telling friends I had taken up with a younger woman named "Rosie Yellow Shirt" (after the yellow Nomex shirts firefighters wear), and was just using fire as an excuse to return to the reservations.

CHAPTER 34

1985

We were in an impressive drought cycle in much of Montana. 1985 promised to be another dry year.

Our first call was back to the Northern Cheyenne. This time, it was the Eagle Cry Fire, just east of Crazy Head. We set up our base camp at Crazy Head Springs again. Eagle Cry wasn't that tough; it burned through some of the roughest country on the reservation but we caught it as it burned onto flatter terrain.

Next call was to the Wheatwell Fire along Pumpkin Creek on the Custer National Forest out of Ashland, Montana with fuels similar to those on the reservations. The country was quite broken, with a lot of "badland" terrain. The fire started when someone set fire to a garbage dump on private land. It spread rapidly to the east, driven by high winds that created a long, narrow fire aligned from west to east. The fire was probably about half a mile wide and about five miles long. It burned across an interesting mixture of lands administered by the U.S. Forest Service and private land owned by several individuals.

Most of the private land was posted with no trespassing signs with locked gates on every road. We had a tough time finding owners to get keys to gates and permission to cross, although the fire was burning up a lot of their livestock forage and fences. Actually, most of the ranches were owned by wealthy "absentee" owners who depended more on the tax advantages involved rather than the commercial production from the land. One owner refused to let us use water from his pond because he thought we might scoop up a fish. A lot of his property burned.

Numerous other fires were burning at that time, especially in Montana. One, in the Gates of the Mountains area near Helena was especially troublesome. Supplies were hard to come by.

We caught the head of our fire in a field of recently harvested winter wheat. Local initial attack forces had gotten most of the necessary line in place on the north flank so we could tie off the head of the fire by the end of the first day. The fire wasn't moving to the south, so we ignored that flank.

The badland formation intrigued me. Erosion has exposed layers of sedimentary formations. A lot of coal has developed in the area over geologic time. Many of these coal seams have burned in previous wild fires, leaving layers of red scoria, interspersed with sedimentary layers.

Several coal seams were ignited by the Wheatwell Fire. We initially gave more attention to the coal seams than completing the line along the south flank. An ominous weather forecast immediately changed that priority on the third morning. We received a "red flag" weather forecast about eight a.m. A cold front would reach the area about mid-afternoon. I immediately called everyone off other assignments and threw everything we had into that south flank. I got a lot of questions about that strategy. Everyone had been concentrating on the coal seams and that still seemed to be the highest priority.

We experienced strong southwest winds on the fire up to that point. A cold front wasn't going to drop the temperature that much, but it would bring strong winds that would shift to the northwest. A northwest wind would drive full force against the southern flank we had been neglecting. I really didn't want the fire to switch from a half mile front (that we had cut off) to a five mile open front on the south side. We built a lot of line and put out a lot of fire in the time we had left.

I was extremely pleased with the progress we made. I was tracking the advance of the cold front whenever information could be picked up on my radio. Billings Dispatch indicated that the front arrived there about 3:30 p.m.

I was walking the south flank about four. The air was strangely calm for mid-afternoon. A lot of "dust devil" whirl winds had plagued the fire all day, indicating unstable air. Suddenly the air was filled with flocks of birds, all heading towards some dense trees along Pumpkin Creek. Clouds of dust and ash began to materialize inside the burned over area to the north; little bursts at first, then one gigantic cloud.

There really wasn't any place to go. We just stood there, mesmerized by the spectacle. And then the front arrived. A couple of sudden wind gusts hit us at first, then, bam, the wind was so strong that we had trouble standing. We didn't want to stand since the air was full of dust, ash and smoke all moving so fast it stung our faces and made breathing difficult. We hunkered down behind a rock ledge, waiting for a break which took a couple of hours to arrive.

We had made the right move. Thousands of acres burned elsewhere in Montana that day while our line held. It took us several days more to put the coal seams out, always tough mop up. Although we were sure we got everything, I understand that at least one of the seams started to smoke later that fall, and needed more attention. A coal seam resembles a charcoal briquette: leave one little spark, and it will eventually spread to consume the whole thing.

We didn't get called out again as a team. The "East Side" Type Two Overhead Teams were set up so one person can't occupy a specific position for more than three years, the intent being to rotate qualified personnel from various agencies through the few available positions. My three years as Incident Commander were up. I could have rotated into another position, such as Operations Section Chief, but opted to go on the alternate list instead. I lost a key range management assistant through budget cuts and simply had too many other commitments that required additional attention in my staff officer position.

Besides, my being tied up on the teams interfered with time I really needed to spend with our family. The kids were growing up fast and would be leaving home all too soon. There were other conflicts.

Dillon has a big rodeo/county fair over Labor Day weekend. Our children were active in 4-H and Lois and I served as leaders in some options. For two years in a row I started to help our youngest, Teri, get her horse ready for the 4-H horse shows. Our horses had been trained to climb mountains and pack camping gear. They were not used to performing in an arena atmosphere, especially when receiving a bath from a garden hose was a prerequisite. Some of the first baths tore up the better part of an acre before the horse was deemed clean enough for horse shows. We'd get her horse about half bathed and I would get a call sending me off somewhere to a fire. Lois had to leave her busy job to help Teri while I answered the fire call. Lois remained hopeful that I could give the horse a complete bath at least once. She emphasized that point on at least one occasion by using my electric razor to fine tune the horse's trim. I don't recommend that procedure if a man intends to continue using the razor.

CHAPTER 35

On the Alternate List

My normal job demands were increasing. Environmental groups were becoming increasingly concerned about livestock use on public rangelands, especially along riparian areas. Ranchers tended to resist change because of the increased expenses involved. Most family ranches were running on a very fine line economically, so their concern was very real.

The drought continued into 1986. We were also completing a land management plan that addressed all uses for the whole forest. I could spend chapters pursuing the political dilemma facing the Forest Service during this period, but will try to sum the situation up in a paragraph. Preservation interests were using emotional issues to promote their personal businesses by appealing and/or litigating every effort made to manage the multiple-use resources available on the nation's forests, regardless of environmental consequences. Industrial interests were lobbying congress and appealing/litigating every decision to reduce production. The result was conflicting legislation from Congress and a maze of legal mandates favoring emotional rather than scientific management. The Forest Service lost identifiable missions and directions during this period.

I had to turn down several fire assignments on various overhead teams. I really wanted to go since we could still fight fire, but there was so much to do and not enough of us to do it. I was getting frustrated. Fires continued to burn to the point that almost all qualified personnel were on the fire line.

Dispatch made their normal two a.m. call one night. Maybe they think a firefighter isn't as likely to turn them down at that hour. The Lolo National Forest had a problem fire burning just north of Missoula. All of the Type Two Teams were already on fires. They were trying to put together an additional team off the alternate roster to take over this one. Would I go as Incident Commander?

I was on a plane en route to the Houle Creek Fire on the Nine Mile Ranger District at daylight. I had been on fires with most of the alternates lined up for the new team and I was confident we could do what was needed.

The Lolo had placed some jumpers on the fire initially. They declared it out and left. The fire took off again. The most embarrassing thing about this fire was that the smoke was visible from the Forest Service Regional Office in Missoula. It was burning just above major subdivisions that included some very expensive homes. The Regional Forester wanted it out completely as soon as possible.

Most effective resources were already committed elsewhere, but we managed to pull some in. The visibility of the fire didn't hurt. I doubt that we could have gotten a good helicopter with water dropping capability, along with numerous retardant drops from the nearby fire center if politics hadn't been involved.

The new district ranger had no fire experience. She had been a fisheries biologist back east before being promoted to district ranger in Montana. I assume she was a very capable fisheries biologist although personal observation indicated the system had pushed her too hard. With all of the priority on affirmative action, the government pushed a lot of otherwise capable minority/female individuals in over their head before they learned to swim and then wondered why they failed.

We were finalizing plans in the parking lot at the ranger station for our initial action when one of the ranger's two huge dogs lifted his leg on my fire pack. I responded by kicking the dog where it hurts the most. I don't think it had ever been kicked before. That caused a few delays.

The fire wasn't the toughest although it was burning in very heavy fuels on steep slopes just above numerous very nice homes. Our first priority was to ensure that it wouldn't reach the subdivisions. We got several Montana Indian Fire Fighter crews and hit the line. We got a pretty good handle on the fire the first day. Mop up was going to be a major problem in the heavy timber. The helicopter with its water bucket was important to wrapping things up in a reasonable period.

I flew up to inspect the line on the third day then sent the helicopter to drop water on hot spots. Things looked pretty good although the weather was hot, dry and windy. I ate lunch with the Operations Section Chief on an open rocky point where we could look out over Missoula.The big "M" on Mount Sentinel overlooking the University of

Montana was visible just beyond Missoula, in spite of the smoke from numerous forest fires to the west. For some reason, someone manages to set Mount Sentinel on fire every couple of years. We joked about it being time for it to burn again.

I made one more hike around the fire before flying off so I could get back to camp in time for a planning meeting. We weren't quite back to the helispot when I noticed that the helicopter hadn't made a water drop for some time. I radioed the pilot to see what was going on. He responded that the fire center had called him on another frequency. He was en route to a major fire on Mount Sentinel.

The Operations Section Chief and I went to the opening where we could see beyond the trees. It looked like everything southeast of Missoula was on fire. An ex-smokejumper with some mental problems apparently was disappointed that he was missing out on all the excitement and started the fire in subdivisions next to Mount Sentinel.

I wasn't about to lose anything with that major blaze pulling resources so I stayed on the line. Everything held. I hiked off the line ahead of the crews that evening to hold the belated meeting. Crews move at the rate of the slowest member so it takes some time to work them off steep slopes. Some other overhead and I reached the bottom well ahead of the crews. We had to walk through the back yards of some pretty fancy houses before we reached the turnaround for the buses where a truck waited for us. A little old lady ran out of one of the houses. She had a pitcher of iced tea and homemade cookies.

"I just wanted to show my appreciation to you fire fighters," she said. "I want to give you all something special when you come off such a hard day's work!"

That tea and cookies looked mighty good. I assured her she was just the sweetest thing, and thanked her for the thought. Usually we only hear from the public when they think we have screwed up.

"You do know that there's about a hundred other fire fighters behind us, don't you?" I advised her.

"Oh my, no!" she responded. "I don't have enough tea and cookies for that many."

She thought for an instant, then said "Why don't you people drink up this pitcher, then I'll just disappear."

She was as smart as she was sweet. I think that was about the best iced tea I ever tasted and the cookies were chocolate chip, my favorite. I tried to feel guilty.

It took us several days to wrap things up without the helicopter before heading home to see what developed next.

CHAPTER 36

Bad Times and New Assignments

Some special personal problems developed in early August, 1986. Lois and I took our horses out for a run one Saturday. I wanted to clean up a littered hunting camp I located on an official ride with others earlier in the week.

My saddle horse got the lead rope from a pack horse under his tail. I had no idea that horse could buck like he did! I rode him, but ruptured a disc in my lower back in the process. Related sciatic pain kept me flat on the floor for the rest of the fire season.

I finally gave in to the pain and went in for surgery the following February. I obviously didn't ask the right questions. We went to a hospital in Butte, Montana for the surgery. Since then, I have been advised by others that they have an unusually high incidence of secondary infections following surgery. I wish someone had warned me earlier.

I think the surgery was a success but somewhere along the line I contacted the staph infection that causes toxic shock. The result is a story of its own. I died, had a very personal discussion with "The Light," and He let me come back. My kidneys and liver failed and I was evacuated from Butte to Missoula via life flight helicopter for treatment. I spent thirty-two days in intensive care in Missoula but survived with some very painful memories. My recovery took a long time.

Fires took off again in 1987. It wasn't long before much of the west was on fire, especially in Washington, Oregon and California. The fire folks were justifiably concerned about putting me back on the line, but they needed help. I headed for Medford, Oregon as a Regional Crew Representative, a significant switch from normal fire line duties. Now my job was to monitor crew location and assignment, monitor safety and morale, and handle any problems that came up involving the crews. I'm not sure I make the best baby sitter. I had a fancy rental car to drive and got to stay in air conditioned motels.

Things went pretty well in Oregon. The whole western part of the state was socked in with smoke, but crews were making progress on the line. The most common problem I encountered related to use of marijuana by crew members.

Montana is cold country; most of it too cold to grow good marijuana. Oregon and California aren't. Lots of enterprising entrepreneurs look at isolated tracts of National Forest and associated lands as an excellent place to cultivate marijuana. If authorities find it, it isn't on the grower's land so they can just walk away and don't get caught. Lots of marijuana burned along with other fuels in those fires. Unfortunately, some of the firefighters recognized an opportunity. They would find some marijuana growing along the line, do a little harvesting of their own, and inevitably smoked a bit. The dumb ones got caught and we had a lot of dumb ones. Drug laws are more lax in Oregon and California than in Montana, so mostly we just sent them home.

I eventually transferred from Medford to Redding, California, where the region had more crews on bigger and hotter fires. I didn't get close to the fire line. We had an office set up for crew representatives adjacent to a retardant base. We couldn't stay off the phone-I spent twelve to fourteen hours a day in that hot, cramped office talking on the phone and filling out forms.

Thousands of firefighters from all over the United States were in the West. Relatives died, wives had babies, spouses filed for divorce (not an unusual problem for firefighters), family members got in car accidents, etc. We had trouble hearing over the phone every time a plane came in for another load, but we did the best we could. We'd get a call then try to track the person down to relay the message, and, when necessary, make arrangements to ship them home. We also had to work the other end. Whenever someone got hurt or in trouble we had to call their home base. Drug problems continued.

One of the biggest headaches developed when an MIF Crew from northeastern Montana got paid before they left a State of California fire, followed by a day off. Don't do that. We normally hold payment for all Type Two Crews until they get home. Predictably, this crew immediately went on a wild drunken party and got into all sorts of trouble.

An enterprising member collected money from fellow crew members and went absent without leave, looking for the best drug deal he could make. He was going to take the drugs back to the reservation where he could make lots of money and share the profit with his buddies.

The state police found his basically headless body under a bridge. The drug connection had everyone side tracked for a while. It eventually surfaced that he never got that far. Two fourteen year old boys went on a killing spree. They started in southern California where they murdered a young woman and stole her car. They proceeded up the coast, blowing the heads off folks they encountered along the way with a 12-gage shotgun. They found the firefighter thumbing his way down the highway in quest of drugs and saved him the trouble. We had to fly his family from Montana to California to identify the headless body and return it to Montana.

Type Two Crews were normally assigned for fourteen days then returned home. Overhead and Type One (hot shot) crews were assigned for up to twenty-one days, with a day off on the fourteenth day to rest. Firefighting is some of the toughest work there is, so personnel get quite worn down with a couple of weeks on the line. Some of the crews came a long way, including crews from eastern states. It was expensive to transport them to the fire line, work them for a couple of weeks and then have to send them home. They'd just rest there for a day or two, then report back to some agency and volunteer to go out again. The taxpayer would have to finance the process all over again.

The decision was made to try a rest and recreation program (R&R), similar to what is provided soldiers during war time, where we moved folks into special camps to rest up instead of sending them home. Resorts were contracted, or people were put in tent camps in public parks with facilities such as swimming pools and tennis courts.

It worked marginally. I sympathize with those responsible for R&R camps for combat troops. Humans can find innovative ways to get into trouble. Some of the camp officers got tired of having to bail people out of jail after they tore up a bar somewhere and restricted some crews and individuals to camp. They were immediately hit with an avalanche of charges of civil rights violations. It was a lively time.

One R&R camp was in a place called Anderson Park. The occupants immediately dubbed it "Andersonville" after an infamous Civil War prison camp. They apparently had a very strict camp staff.

We eventually put in our time and were rotated out. I was happy to leave. In retrospect, I'm sure that having to sit in an office chair glued to the telephone was much tougher on my back than a decent job on the fire line ever could have been.

Photo Gallery

SUPPLY DROP: Food, water and other much needed supplies arrive on Hoodoo Ridge.

COLLIE LAKE FIRE: A helicopter view while directing smokejumpers and retardant drops. Note the abundance of spot fires.

MIDDLE FORK CROSSING: Ferrying fire crews across Idaho's Middle Fork of the Salmon River at the mouth of Greyhound Creek.

TOO HOT: A firefighter retreats from a hot spot on Mortar Creek.

BURNOUT ON GREYHOUND: Burning out the fire line on Greyhound Creek. The two bright spots are backfire fusees.

TOO HIGH: A plane starts a retardant drop. Topography and increased winds prevented the plane from getting low enough to help.

RETREAT TO H-1: Spot fires surrounded us as we retreated to our camp and related personnel facing a burnover by the Mortar Creek Fire.

H-1 AT RISK: We started a backfire to counter the approaching crown fire as I took this photo. The rock pile in the foreground appears in the next picture.

BURNOVER AT H-1: Our backfire joins the main blaze as the crown fire sweeps over the camp. Reference the rock pile in the foreground in the previous photo.

MCCARTNEY MOUNTAIN: An aerial view of one of the fires I picked up while serving as an aerial observer in 2000.

CHAPTER 37

1988

1988 made the media as the year the West burned. We had very little snow the previous winter, and even less rain that spring. Fires started early and continued until snow finally arrived late in the fall.

The fire organization still had reservations about sending me out in a line position. Things kept getting worse and worse, and they were running out of qualified people for badly needed overhead positions.

Major fires were burning all over the west. Firefighting resources were in short supply and it was becoming impossible to keep track of all the manpower and equipment assigned to various fires being fought by various agencies. An Interagency Coordination Center was established at the Jumper Base in Missoula. An Interagency Coordination Center tries to ignore real and political differences between various national, state and local firefighting organizations while ranking existing and newly reported fires by priority based on life and property at risk. When resources such as personnel and equipment become available they are sent to the fires having the highest priority. The centers also attempt to ensure assigned personnel work safely and get rotated into rest centers or sent back to their home stations as required; keep the media, politicians and representatives from all concerned agencies briefed daily or as requested; and handle anything else that might develop.

Personnel in the Missoula Interagency Coordination Center became concerned about progress being made on the Northern Cheyenne Reservation where a Type Two Team was assigned to handle a complex including several fires.

The Missoula Center feared that the team was in over its head considering current and predicted weather conditions and the lack of additional personnel and equipment to meet everyone's needs. They asked me to go in to see what was really going on.

I could see several large convection columns from different fires as I drove towards Lame Deer. Burning conditions continued to deteriorate every day during July through September of 1988. It was apparent that we really weren't going to be able to catch everything until we got a major weather break.

The reservation fires had burned right up against Lame Deer, reaching several abandoned vehicles on the edge of town before being controlled. Considering burning conditions, the Type Two Team was doing a commendable job and there was no need to replace them.

I called that information in to Missoula. Unfortunately, they had gotten too nervous to wait. They ordered a Type One Team out of Alaska when it became available and it was too late to turn them back. I hate second guessing by managers who don't know what is really happening on the ground. I advised them they had made a mistake and were wasting thousands of tax payer's dollars. I'm sure they needed that Type One Team somewhere else a lot worse, but I was too late. The wheels were in motion and the new team would arrive in a couple of days.

I did what I could to help out at Lame Deer for a couple of days until the Type One Team arrived, then headed for Billings for the night. Personnel from the coordination center phoned me at the motel. They didn't feel they were getting needed information from another Type Two Team on a fire on the Rocky Boy Reservation south of Havre, Montana. They wanted me to see what was going on there while I was in Eastern Montana.

I found the team firmly in control at Rocky Boy. They were having problems getting needed supplies, which is normal for a Type Two Team when a lot of major fires are burning. They get used to making do, and this team was doing great.

The fire was burning adjacent to an area that had burned in a major fire a couple of years before. They were herding the current fire into the old burn, a sound strategy. Perennial vegetation normally will not burn again for several years following a fire, so recent burns make effective fire breaks.

The Coordination Center waited for my report this time. I doubt that they had another team to send anyway. I spent a couple of days on the Rocky Boy then returned to Dillon.

I hadn't been home long before the next call came. This time I was asked to take over the lead position as

Planning Section Chief at the Interagency Coordination Center in Missoula. In this instance, the Planning Section Chief supervised the coordination operation.

It's an impossible job, especially on a year such as 1988, with over 20,000 firefighters, every retardant plane in the U.S. (and some borrowed from Canada), hundreds of helicopters and bull dozers and every other type of firefighting equipment available fighting over a hundred major wildfires in the western United States.

The National Guard had been activated in several states, and eventually several active military units were sent to the fire line to help. There was enough "brass" from all of the agencies involved to fill a good sized conference room for every briefing. There was a lot of resource values at risk, burning conditions were about as serious as they could get, and no break was in sight. We settled in for a long haul, and did the best we could.

Each day started at five a.m. Our team in the coordination center included specialists in public information, fire behavior, weather, logistics, finance, safety, communications, personnel and affirmative action. Each day we reviewed all information available on new and on-going fires including the values at risk, predicted weather, resources assigned, and requests for new crews and equipment for each fire. We based daily and projected priorities on this information. Next we identified any new or re-assignable resources that might be available to respond to requests. We'd use our established priorities as the basis for distributing the limited personnel and equipment as it became available.

We used this information to develop a daily plan of operations by 10 a.m. each morning, when we conducted an agency and media briefing. We developed charts, graphs and other information before each briefing, and made copies concerning fire priorities, size, values at risk, etc. National TV representatives started setting up about 9:30 a.m., followed by lots of other very important people. We eventually had to limit attendance to fit everyone into the available room.

The briefing was supposed to last for an hour but routinely lasted two. We'd introduce the people representing the Forest Service Regional Forester, State BLM Director, Bureau of Indian Affairs and tribal representatives, Director for State Lands or their representatives from involved states, politicians, top officers for the National Guard and active military units involved, and other concerned officials attending each meeting. Montana's governor was very concerned and attended several briefings in person.

After introductions, we'd brief everyone on the existing conditions which continued to deteriorate day by day. Fires were burning everywhere, requests for personnel and equipment far exceeded what was available, and the weather continued to get hotter and dryer. We simply did not have enough resources to meet demand. Some of the fires were so big that they were going to burn what they wanted to burn in spite of anyone's best efforts, a point that affected priorities.

Each specialist gave a presentation at the briefings, discussing what they predicted for the future. We were doing amazingly well on safety, considering the number of people assigned and the nature of their assignments. The finance folks came on last. Fire control efforts were eating up millions of dollars a day. We closed by allowing the officials to make statements for their interest (generally political, and frequently demanding impossible things) then we took a fifteen minute lunch break.

Each afternoon was spent monitoring any information we could get from various fires. Fire burns best in the heat of the day. They continued to burn extremely well every afternoon. This information was used to prepare another plan for the night, and another briefing at six p.m. The afternoon briefing format essentially duplicated the morning session, with current information added. After the briefings, we'd update all information we could accumulate for the following day, and generally signed off about ten p.m. Then it was out to a quick supper and a motel for a few hours of sleep with the routine starting again at five a.m. the next day.

Predictably, the media got hung up on Yellowstone National Park since it is a well known landmark. The fires burning there were long overdue. Fire had been kept out of fuels that routinely burn for all too many years and nature was just catching up. Granted, some man made features such as major lodges were threatened, but we could assign people to protect them. Yellowstone wasn't our top priority. Whole towns were either burning or were threatened elsewhere; major commercial timber stands were burning; ranches, resorts and other structures were either on fire or in the path of fire.

We had to give Yellowstone some attention, primarily because of the visibility created by the media. It also represented a "black hole" for personnel, equipment and supplies since anything sent there simply disappeared into an internal organization. There was no way to relocate it for reassignment when other more pressing priorities developed.

A lot of local fire departments have agreements with other agencies to make their equipment available when it is not needed at their home base. Not only does that option add significantly to the national inventory of available equipment, but local fire departments gain excellent experience and receive good pay for personnel and equipment when they are dispatched. This money helps local fire departments purchase needed equipment. Predictably, we kept getting hysterical calls from people whose homes were threatened, demanding to know where their fire equipment and firefighters had been sent. They paid taxes to maintain local fire departments, and now they needed them at home to keep their homes from burning.

For example, a major blaze started near Lolo, Montana, when some idiot, ignoring all warnings sent out through the media and personal contacts and the very obvious fire danger, decided to burn some garbage in his back yard. Lolo and eventually a lot of subdivisions around Missoula were threatened by his fire, and some structures burned. Most local fire equipment was assigned elsewhere, most of it in Yellowstone where it had simply disappeared. People losing property in that fire chose to blame the government rather than whoever started the fire. The government has more money.

We didn't envy firefighters assigned to the more remote fires. Essentially every piece of firefighting equipment was assigned on the fire line somewhere. We had an impressive pile of back orders for personnel and equipment that simply did not exist. These requests just kept piling up. Whenever we did manage to break anything free, we had to send it to the priority fires where major towns and other resource values were immediately threatened.

A major blaze took off in the Madison Range on the

Beaverhead National Forest. It was burning through unmerchantable timber, grass and sagebrush on the west side of the Madison Range. It didn't pose a major threat to anything so we had to assign it one of our lowest priorities. The smoke column was visible from Big Sky Resort between Bozeman and West Yellowstone, and a lot of expensive homes are located in that vicinity. The fire had to burn a long ways north before it could get around miles of solid rock at timber line to reach those expensive developments.

Folks were getting pretty paranoid. The media was doing everything they could to encourage the paranoia by sensationalizing everything, which definitely did not help. It was obvious to some of the owners of the expensive Big Sky homes that no one was doing anything. The media had several fat ladies and a drunk man on TV that night, crying and carrying on about how the fire was going to come roaring over the rocks to destroy everything they held sacred, and nobody was doing anything to stop it. The media really had to look to find those clowns. They tried even harder to give them credibility. Everything was the government's fault.

Had the fat ladies, drunk and reporters been in better condition, I'm sure the folks on the Beaverhead National Forest would have welcomed their help instead of criticism. About all we could send the firefighters on that blaze were a handful of smoke jumpers and a twenty person Type One Hot Shot Crew we scrounged up somewhere.

The few people left on the Beaverhead National Forest, Madison County's fire departments and other local agencies, plus the limited resources we sent deserved medals for the job they did on that fire with their limited resources.

Other fire fighters did similar things to control other fires. The media somehow overlooked these successes and chose to look elsewhere for sensationalism and people who were more than anxious to criticize the firefighting effort.

Other fires kept cropping up, more than enough to outpace the few that crews managed to control. The major blazes weren't about to slow up until the weather changed.

Agencies set up major training efforts around the country to recruit new crews. Thankfully, a lot of this effort was directed towards the military. Military personnel are generally in reasonably good physical condition, are used to taking orders, and know how to work as a team so they made pretty fair fire crews once they were trained in fire control. It would help to give them more experience in living in tent camp conditions in mountainous terrain. Few of the military personnel knew anything about working in the mountains, but not a lot of people do.

Other efforts were directed towards recruiting crews out of cities such as Butte, Spokane, and even some eastern cities. In general, those from the east didn't understand the mountains, but were more likely to try to work. Essentially everyone who didn't have a job and actually wanted to work in the west was already on the fire line. We were down to the bottom of the welfare barrel. The remaining unemployed in reasonable physical condition understood the welfare system a lot better than physical labor in hot smoky places in the mountains.

Then there were the complaints. If a home was threatened (lots were), and the owners didn't think we were doing enough to save it, their call generally got routed through to us. We also got a lot of calls from unhappy folks who had contracted to send equipment and supplies to the line.

Contractors were making a bundle off the situation. There just wasn't enough equipment to meet demand, so equipment owners could pretty well name their price. There were minimum standards that had to be met, but agencies were short on inspectors and long on need. A lot of worn out bulldozers and other equipment hit the line to earn their owners enough money to buy new ones.

Large buses are critical to get crews from fire camps to the fire line. Agencies have agreements with many school districts to use school buses. Unfortunately, schools start up in late August. Fire season extends well past the time the buses go back to transporting kids.

A company in Bozeman owns a lot of commercial buses. We had them all contracted, and on the fire line somewhere. I eventually got on a first name basis with the company owner. He wasn't complaining about what he got for renting the buses. His problem was in finding qualified people to drive the buses for union scale wages. He would find a qualified person and get them in a bus on the fire line somewhere. The drivers got the crews on the line then didn't have anything to do but sit around until it was time to bring the crews back to camp.

The drivers would talk with the agency logistics and finance personnel back in camp during their spare time. These people were facing real problems in recruiting enough qualified individuals to run motor pools, keep time on rental equipment and investigate damage claims. These agency jobs paid better than the union scale wages the owner paid the drivers.

The owner of the bus line was uptight about his doing

all of the work to find qualified people and get them out there, only to have agency folks steal them from him for higher pay. He had a good point but all I could do was sympathize. I requested that the agency people quit stealing the drivers, but to no avail.

One ten a.m. briefing ended in a media stampede. We received word just as everyone was entering the room that we had at least one fatality on a fire at Polebridge, Montana. All we could find out was that a night crew had gathered where they were to meet a bus to take them back to camp at eight a.m. While they were waiting, a large snag (dead standing tree) fell right in the middle of them. We had no idea concerning critical details, such as the crew involved, how many people were killed or injured, identity of next of kin, and other critical information necessary before we could go public. The public information specialist asked what we should do. I suggested that our best bet was to wait until we got a reasonable amount of information then make a news release as soon as we knew what actually happened. It was too early now.

Over 20,000 people from all over the United States, and some from Canada, were on the fire line, or at least in fire camps. The relatives back home had no idea where their loved ones were or what they were doing. An announcement on national news that a firefighter had been killed without any additional information would trigger panic in the hearts of anyone with a loved one in the west. If word got out prematurely that someone had been killed, all involved agencies nationwide, and especially our unit, would be totally paralyzed by phone calls until we could get the information needed, notify the next of kin, and advise the media.

We assigned team members to each door to brief all agency personnel who might have gotten the word, requesting that they say nothing until we knew more information. We thought we contacted everyone. The briefing appeared to be going great. We went through all of the specialists up to the point where the Finance Chief gave her presentation to summarize things and I thought we had it made. Then some idiot from the Forest Service regional finance office, who had somehow slipped into the briefing but wasn't scheduled to say anything, stood up and announced that a tree fell on a crew at Polebridger and they anticipated financial claims. He treated it like he knew something we didn't, and made a big deal of it. The media went crazy. I could have strangled him, especially with his superior attitude. He didn't have a clue about the problems he had just created.

We spent the rest of the day and most of the night answering calls from at least 20,000 concerned mothers and other relatives. We finally found out that the man killed and people injured were from the Coeur d' Alene Reservation in northern Idaho. We got the next of kin notified, made the appropriate news release, and were able to get back to normal duties a few days later.

The media made a big deal out of the "let burn" policy some agencies were employing. These agencies, including both the Park Service and Forest Service, had finally recognize the role natural fire plays in maintaining environmental conditions over time. They developed necessary fire management plans for specific areas, usually parks and wilderness areas, to allow some fires to burn naturally as long as critical values aren't threatened and the weather meets certain criteria.

They implemented some of these plans in some areas early in what had been predicted as one of the worst fire seasons on record. The plans should not have been implemented that early in the season. We exceeded weather condition limits identified in the plans by early July, and were now into late August. The unfortunate point was that conditions got so bad that several of these "managed" fires had exceeded planned limits weeks before, and were now as much of a problem as others that we had tried to control from the minute they were detected. Letting those fires burn in June when predictions indicated serious burning conditions in following months was a serious mistake.

Mom was right. You will get burned if you play with fire. Some days, especially the infamous "red flag" days, were worse than others. A cold front or other weather feature would be predicted, assuring high winds, low humidity and major fire spread. The red flag forecasts are designed to advise line personnel to expect extreme fire behavior. Fire crews did what they could to hold under the predicted conditions. There really wasn't that much that they could do when they encountered that much fire in those fuels and those conditions.

For example, the Canyon Creek Fire had started within conditions specified in one of the fire management plans in the Bob Marshall Wilderness in Montana. Weather changed until conditions were well beyond planned specifications so crews were trying to stop the fire. Since it was a wilderness fire, we couldn't give it the priority we did fires near towns, but forces assigned there were doing everything they could to stop the fire from advancing with the resources we could get them.

A serious red flag forecast was predicted. The Canyon

Creek crews knew that their only hope was to try to hold the fire along a major ridge near the wilderness boundary. The predicted front hit and the fire went over the line people like they weren't there. Some crews had to deploy shelters. It wouldn't have mattered if there had been twice as many fire fighters and equipment on that fire at that point. The fire burned clear out of the Rocky Mountains and made a major run across the plains for miles towards the Missouri River south of Great Falls, Montana.

Thankfully, Canyon Creek didn't burn up any towns or people. It consumed several ranch buildings, haystacks and crops. It's amazing how valuable hay, fences, run down shacks and a few more modern houses become when they burn. Ambulance chasing lawyers are more than anxious to help those affected sue the government. I'm sure the taxpayers are still paying on that one. A lot of other fires went crazy that day too.

Montana's upland game bird, special big game hunts, and archery hunting season were rapidly approaching. We had a lot of concerns about the problems a major flush of people in the woods would cause, both from the opportunity for new starts and the safety aspects of having a lot of untrained people wandering around near major wildfires. Montana's governor was especially concerned. Hunting represents big business in Montana. A lot of Montana industry (sporting goods stores, outfitters, guides, motels, gas stations, lodges, etc.) depends on hunting for a large part of their income.

The governor flew to Missoula from Helena to attend one of our briefings before making a decision. It was a major media event. Our Fire Behavior Specialist reviewed all of the information he could get. The outlook wasn't good. Based on his calculations, any ignition source, be it a dropped cigarette, match , automobile catalytic converter left parked in tall grass, etc., was over ninety percent certain to ignite any vegetation it contacted under existing burning conditions. The governor postponed hunting season in Montana.

I had been in Missoula for close to four weeks, and the pressure was taking its toll. I had taken a couple of hours off to accompany Lois to a neurologist in late August. She had been experiencing some unexplainable physical problems for several months. Our doctor eventually suggested that she might have Parkinson's disease. The neurologist confirmed the diagnosis, a fact that hit us both hard. We couldn't change the diagnosis or the disease. I'm sure it wasn't fair to Lois, but the fire assignment helped me cope with the diagnosis, so I remained on in Missoula.

I received a couple of days off shortly after Labor Day. Idaho's governor hadn't postponed hunting season there, and my brother, Lew, had a permit to hunt bighorn sheep in the highest mountain range in Idaho. I went hunting bighorn sheep.

It was a tough hunt but a badly needed break. Smoke seriously limited visibility, and the dry conditions and heat had animals where we normally wouldn't find them. I climbed a lot of rugged mountains over 11,000 feet in elevation in the Lost River Mountains. Then it was back to Missoula for another couple of weeks.

The second tour wasn't much better. Essentially all fire crews and equipment were getting run down. Some of our major problems now related to responding to claims for worn out equipment, and damages and news events caused by personnel when they were given days off. I'm amazed at the variety of trouble people can get into when they get a break after a long spell of hard, hazardous duty. Lodges in Yellowstone were essentially closed down because of the fires. We used several of them as R&R centers for fire crews. Crews got drunk and tore things up.

I remember a very irate letter to the editor in the media from a lady who had scheduled a family vacation. They were given the option of a refund or to go ahead and brave the fires and related activity. The family came to Yellowstone. They couldn't explore the park since essentially all roads were closed because of the fires. They couldn't enjoy the lodge because it was overrun by drunken firefighters. Their vacation was ruined. It was the government's fault.

An integrated fire crew (both males and females) hit Bozeman. They immediately went to a local Laundromat to wash clothes. They chose to bring along several cases of beer and wine. Then they decided the clothes they were wearing really weren't that clean either. The police found twenty naked drunk firefighters in the Laundromat when they arrived. At least I assume they had clean clothes on when they went to jail.

R&R can be hazardous to your health. A young lady survived twenty-one days on the fire line in Yellowstone, then got so drunk she fell off a sidewalk in Bozeman and broke her arm. Two members of a Snake River Valley crew got drunk, got in a knife fight, and ended up in the Bozeman hospital.

Owners from all over the West were calling to make claims for damages done in motels and lodges. I am amazed at how valuable a K-Mart special picture on a motel room wall becomes when someone damages it. R&R didn't cost the taxpayers as much as fighting fires, but it was significant.

I was happy when my tour ended so I could return to the Beaverhead to see how resources under my charge were holding up to the drought and fires.

The fall snows were late in coming. We had personnel on the fire line until early November. We were happy to see winter and the end of fire season.

CHAPTER 38

Back on the Alternate List

Essentially all management activities on the Beaverhead National Forest were tied up in appeals, law suits and various other bickering between environmental interests and industry. I lost a range conservationist and a wildlife biologist who had furnished me critical assistance. I couldn't replace them because of reduced budgets and had to pick up the slack myself.

Most of the ranchers I worked with wanted to do the right things; they just had to do it in a manner that allowed them to make sufficient profit so they could stay in business. The economics of land management is a very important point that frequently is overlooked. The Forest Service produced more revenue from management activities than Congress allocated for operations for most of its existence. By the mid-80's the chain of expensive environmental documents needed before any activity could be proposed followed by the mass of appeals followed by legal action started driving the cost of any management far beyond Congressional appropriations and potential revenue. We wanted to continue to support local economies through reasonable timber sale programs and livestock use. The alternatives to historic multiple use management included the closure of hundreds of sawmills and unemployment levels that had major economic impacts in hundreds of communities in the west. The ranchers at least had a couple of options: recreational subdivision or sale to rich absentee owners who routinely lock the public out and direct all activities allowed on their property to meet their own selfish desires. Both alternatives adversely affect the general public and environmental needs.

It was tough to find time to worry about fire. Other agency personnel had the same problem. Political concerns were seriously impacting the agencies' ability to get personnel out of the office to address major management needs. There simply was not enough qualified manpower to address all needs. The fires didn't understand and continued to burn.

We remained in a fairly significant drought cycle. Fires continued to escape initial attack in 1989. All of the organized fire overhead teams saw action. I had several opportunities to fill in on teams but had to turn them down because of regular commitments. I eventually got fed up with the lack of progress on needed resource management projects and managed to convince the boss that I needed to go fight fire.

I didn't have to wait long. The Gird Point Fire took off in Montana's Bitterroot National Forest. Bob Michaels was Incident Commander for a darn good Type One Fire Overhead Team in the Forest Service's Northern Region.

They were assigned Gird Point but were short an Operation Section Chief, my favorite position. I find the interaction routinely present between all participants involved in fire control extremely rewarding, but there is something special about the relationship between line personnel. They are the ones at the front where the action and adrenalin rush is. And each person depends on the other members of his team for success.

I got the normal call about two a.m., stumbled around the house gathering up what I needed (with the help of a sleepy spouse who got some coffee brewed and a teenage daughter who filled my canteens), picked up a Forest Service vehicle at the office that wasn't signed out by someone for the next day, and tied in with the overhead team in Hamilton at six a.m.

The Operation Section Chief is responsible for all fire line operations on the fire. A fire team normally has two Operation Section Chiefs. At the time, one was responsible for day shift and one for night shift. Day shift normally experiences the worst burning conditions for fighting the fire, but can at least see what's going on (and has a better chance at getting needed sleep). Night shift is the less desirable of the two. When a person goes out as an alternate to fill behind someone who can't make it, they get night shift.

We sized up the situation on Gird Point. The fire had burned into areas with over a hundred tons per acre of dead trees and other very flammable fuels on the ground. Dead standing snags were everywhere. Slopes were steep, and there were lots of rocks to roll. It didn't take long to decide that a night shift simply was not safe.

The Operation Section Chief position has one major

inherent conflict in meeting all obligations. The incumbent must know what is happening on the ground, continually adjust priorities as needed and control the line operations. Meanwhile, the rest of the team has to plan future actions, radio and other communications needs, order equipment, supplies and personnel, and maintain necessary logistical and financial support. These obligations cannot be met unless the full team understands what the Operations Section Chief needs to support the actual fire control activities. It is impossible for the Operation Section Chief to assess on the ground needs, control suppression efforts, direct line operations, and be at planning meetings at the Incident Command Post at the same time on hot, moving fires.

This overload can develop into major problems in keeping other key team positions current, not to mention building animosity within a fire team. Fires have been lost here.

I had tried, unsuccessfully before, to promote the idea of having three Operation Section Chiefs on Type One teams, one for day, one for night, and one to serve as liaison between what's going on in plans and on the line. And here I was, night Operation Section Chief on a fire without a night shift!

Moving into the coordinating position was a natural shift that worked like a charm. The day Operation Section Chief, Terry Williams, was a good hand who handled the line well. Fire intensity demanded his full attention on the line. He simply did not have time to interact as necessary with the planning section. I could meet him at particularly troublesome spots, maintain constant radio contact with him and review the overall situation from a helicopter. We could agree on needed strategies, and I could make planning meetings without affecting line operations.

Fires are emergencies that tend to generate a lot of internal emergencies as well. I was available to help when we had more than one emergency to handle at a time. And we had some notable emergencies. The weather was hot, and the safety officer continually emphasized the need for line personnel to carry and drink a lot of water. An MIF crew radioed in one hot afternoon that their members were getting sick and passing out.

The symptoms described matched heat exhaustion. I grabbed a helicopter, headed for the nearest helispot and walked up the line to investigate. It turned out that the crew had decided that if water was good, Gator-Aid was better. Those power drinks have some pretty intriguing advertisements, and some people actually believe them. The crew opted to carry a couple of cases of Gator-Aid (free at the commissary) instead of water. Advertisements emphasize that chemicals in the drink stimulate a person's ability to work under stressful conditions but fail to mention that they take water away from your system to do it. We did, indeed, have a whole crew of twenty people suffering from dehydration and heat exhaustion.

We evacuated the worse cases via helicopter while I emptied my canteens into the others, and got more water to those who were still standing as fast as we could fly it to them. Every one survived, but we effectively lost one crew to Gator-Aid.

The extremely heavy fuels and weather demanded application of aerial retardant to help control the fire. The broken topography prohibited much use by fixed-wing "tanker" bombers. They simply couldn't reach the fire in deep canyons. We ordered a couple of heavy helicopters capable of dropping several hundred gallons of retardant at a time. We had to set up a retardant base near the fire line to make this venture effective and reduce costs. We located a wide spot at a road junction on Skalkaho Creek just below the fire, and started to establish the base. Some very heavy equipment is involved. The trucks showed up right on schedule but had trouble negotiating the steep road to the base location. I stuck with them until they started setting everything up then left to attend a planning meeting.

I didn't get far before a Strike Team Leader on site radioed that the road shoulder had collapsed under one of the tank trailers carrying concentrated fire retardant chemicals. The trailer had rolled. Chemicals were spilling from it, right next to the creek.

This wasn't just any creek. Signs were posted all over telling fishermen to release the "sensitive" westslope cutthroat trout they caught. Environmental issues were boiling, with a lot of controversy in the Bitterroot Valley concerning timber harvest and other resource management issues. The Bitterroot is a very polarized valley, full of both environmentalists and loggers. Many of the environmentalists moved there from California, Colorado and other populous areas creating their own environmental problems such as subdivisions, sewage disposal, noxious weeds, demands for more public services, etc. The local long-term residents who have been raised with a totally different lifestyle have a tough time accepting the different values they brought with them.

The Forest Service was caught in the middle and really didn't need a major chemical spill to sully their already questionable image in the eyes of both factions. It's much more socially acceptable to blame the government than on your neighbor, and this would give both sides a common target.

The fire retardant chemical involved had the commercial name of "Phos-chek." Basically the primary ingredients are water, ground up sea weed high in nitrates and phosphates, and a red dye. It actually serves as a fertilizer as well as a fire retardant when diluted with water and dropped on fires.

The concentrated retardant would have diluted eventually in Skalkaho Creek but a major fish kill would develop, probably until the stream reached the Bitterroot River. Having the creek turn blood red didn't sound like a good idea either.

I turned around and raced back to the site as fast as I could. A county road grader was working on the road just below the fire. I radioed for camp personnel to get it headed our way as fast as possible.

The Lolo Hotshots, a pretty savvy crew, radioed me just as soon as I finished that message. They were driving to their assignment on the fire line and were just below the spill. They asked if they could help and I grabbed them; they arrived right behind me.

The chemicals started to congeal shortly after exposure to the air, slowing the speed at which it was flowing towards the creek. The Lolo Hotshots had a dirt dike pretty well in place by the time the road grader arrived. He re-enforced the dike, and no chemicals reached the stream. We later scooped up all of the chemicals and trucked it to an old borrow pit high on the slope that needed fertilizer so it could grow vegetation.

I didn't realize how many people have radio scanners

in the Bitterroot Valley until we had that scare. The Forest Supervisor reported that she had a dozen panic calls from local citizens demanding to know what she was going to do about this major chemical disaster before she could find out what was really going on through official channels.

We were gaining ground, but had a troublesome area that remained to be lined. We also needed to check out a new helispot a crew reported they had cut out on a ridge. The crew thought the new spot could save them a long hike.

I jumped on a helicopter to check things out. I planned to get off at the new helispot, hike down the open piece of line, then catch a ride back to camp with a Division Supervisor when I reached the road in the bottom. I anticipated doing it all before the afternoon planning meeting.

The new helispot looked good from the air. I had a good pilot, and he seemed confident that it was safe. The area had been burned over. We should have been more concerned about a rather obvious hazard. The pilot brought the helicopter in slow, feeling out the new spot. Everything went great until we reached that point where the ship is hovering on a cushion of air between it and the ground. The rotors suddenly sucked up a dense cloud of ash and dust. I couldn't see a thing. The pilot had the choice of trying to fly out blind with burned trees all around us, or continuing our decent. Fortunately, the pilot had flown for commercial ski expeditions, and was used to soft snow landings. The pilot focused his eyes on a stump that showed up through that blinding cloud and brought the ship in for a safe, if very dirty, landing.

We dusted things off; he sized up the best exit route and left me in a very black location on terra firma. I felt a lot more confident there than I had a few moments before. We didn't use that spot again.

The fire ran its course and we caught it with help from a little rain. Having an Operations Section Chief working directly with the planning organization while another directs the line operations worked so well that it was carried foreword to the national level. Some incident teams incorporated the concept into their organization in one form or another. Policy shifts eventually stopped most night operations for safety reasons in the late 1990's.

CHAPTER 39

Six Mile

I returned to appeals, litigation and the job of trying to get needed resource management done on the ground.

I didn't have long to wait to try the planning/line concept for the operations section chief again. 1990 continued hot and dry. Regular programs, essentially all paperwork, kept us pretty busy, but fires continued to burn. The chief of the Forest Service even issued a letter requesting that all but critical personnel be available for fire assignment when all normal fire crews were assigned.

I always got a kick out of that order. Managers who understand fire normally are generous to a fault. They rotate through their personnel so that everyone goes on a fire or two during the fire season, yet have enough people to keep the office doors open at their unit.

Some managers resist this order and go to the other extreme. Their point is that all of their employees are "critical" to something, or they wouldn't have hired them. They refuse to release anyone to fight fire elsewhere.

If these anti-fire line officers are real lucky, they might sneak through a career without a major fire on their unit. They usually experience at least one major "incident" during their career and, thanks to their own shortsightedness, are totally unprepared to deal with the incident when it occurs. They wait until the situation is totally out of control and scream for help. Then they spend all of their time making sure that whoever responds does the job right, although they haven't the slightest idea what that is. While I was with the overhead teams, I tried to identify some trivial item that needed attention, made these idiots think it was important, and got them tied up on that detail while we did our job. These types never learn, and make a career out of being in the way.

We had a major national review involving the Forest's livestock grazing program in August of 1990. I was in charge of organizing the review. I got everything scheduled, then managed to convince the boss that I really needed to go on fires. He agreed, providing I got back in time to lead the review.

The normal middle of the night call sent me en route to the Six Mile Fire on the Gallatin National Forest. The team taking it over was short an Operation Section Chief.

I arrived at the Livingston Ranger Station at about three a.m. to find most of the fire team there but no local forest personnel present to brief us. I sacked out on the lawn until the local folks showed up about five a.m.

The fire was burning in some tough terrain with heavy fuels between Livingston and Gardner. One priority the local personnel identified involved keeping the fire from burning into a wilderness area. The Gallatin National

Forest had enough political problems with wilderness fires in 1988 and wasn't anxious to repeat the experience.

I didn't think that terrain would allow us to commit a night shift, but wanted to review the fire on the ground before making that recommendation. There were no aircraft available immediately, so we headed for the fire in trucks.

We wound our way up a marginal four-wheel drive track to a very nice cabin at the end of the road. The cabin is on an old mining claim that had gone through the process moving it from public to private ownership. There was little evidence that any real mining had ever occurred on the claim.

A young lady and her boyfriend met us there. One of the crew members mentioned that the cabin belonged to somebody named Peckinpaugh. I thought "that's nice," the name meant nothing to me. I assumed he must be some kind of politician since they're normally the only ones with enough influence to gain private title to a mining claim that isn't being used for mining purposes. I never was much good at remembering names anyway.

I heard later that Sam Peckinpaugh was a famous movie producer whose lifestyle led to an early demise. I even think I saw one of his movies once but didn't like it much. I can say that his daughter was a fine host. The cabin site represented about the only level ground in that part of the world. She and the boyfriend volunteered to let us use it for an incident command post and the adjoining grounds for a camp.

The overhead team was made up of good, experienced interagency personnel. I had been on fires with most of them before, and the fellowship and trust was there to make everything run well. The crews hadn't arrived yet, and Norm Fifield, the other Operations Section Chief assigned to the team, was busy ordering what was needed from an operations standpoint. I volunteered to scout the fire to determine what the potential really was.

The boyfriend volunteered to accompany me since he was familiar with the terrain. They were from California, and the elevation slowed him up a bit. He did okay for a flatlander. He loved the area and had a lot of concerns about what the fire would do to the canyon.

I assured him it was a natural process and that nature would fix it over time. I was concerned with what I saw when we reached the fire. The area was overdue to burn with well over a hundred tons of dead fuel per acre on the ground, lots of snags, and plenty of rock to entertain us. The over story was primarily Douglas-fir and sub-alpine fir, with Engelmann Spruce in the canyon bottoms. It looked like a long, hot assignment.

The fire started crowning shortly after we reached the line. Since the boyfriend had no formal fire training, I was concerned about his safety. The fire radio kit hadn't arrived yet, and I couldn't reach anyone from the lower end of the fire with the forest net radio I had. I gave the boyfriend a note describing what we had encountered on the lower end of the fire and sent him back to camp.

I continued around the fire. It was crowning pretty hot near the upper end. I slipped through the fire into an area that had already burned clear to the ridge earlier and climbed to the top through the still smoking remains. It was going to be tough to meet the forest's objective of keeping the fire from burning southeast into the wilderness area.

We had a chance, providing we could get smokejumpers onto the main ridge as soon as possible. I was able to get out on the forest network radio from the ridge. Everyone understood the priority, and the jumpers arrived shortly.

There wasn't that much that I could do there. The jumpers understood their job. I returned to camp. Ground crews were arriving and we had a fire to fight. We agreed that a night shift wasn't feasible. Norm would basically stay in camp, participate in planning and run the air show. I would work the line. I loved it.

Lots of other fires were burning in the northwest. We managed to pick up a couple of good crews who had been rotated back from other fires. Unfortunately, the remaining crews we got were pretty inexperienced.

The jumpers got in a good line to keep the fire away from the wilderness if we patrolled it, so we rotated them out to be available for initial attack on other fires that might occur. We had a pretty secure natural fire line farther east from the fire, although we didn't want the fire to burn that far. The drainage ran into timber line at about 10,000 feet elevation. It was solid rock above that.

The fire burned pretty hot and caused a lot of anticipated problems. Existing fires elsewhere had so drained the manpower pool that some Native American crews were shipped in from Alaska. We received two Alaskan crews.

We needed a couple of crews to hold the line the jumpers had completed on the ridge. The Alaskans were selected based on when they arrived in camp. We could save them a long hike by ferrying them via helicopter to a rocky knob just above timber line. They could walk down a narrow, rocky ridge to the fire line and be relatively rested when they got there.

We loaded them up, flew them to the top of the world, and almost lost all of them. They could hardly stand up, let alone walk. I flew up to see what was the matter.

It turned out that they were from fishing villages in the rain forests along the coast in Southeastern Alaska. To look a long distance up there, they had to look out to sea. Trees got in the way if they tried to look inland. Their idea of a mountain was a few hundred feet above sea level. Rocks were not a common sight. They were loaded on an airplane (a first for most of them); spent a couple of days en route; then they had their first helicopter ride, and we dumped them out on a rock pile at about 10,500 feet elevation. They could see a lot more of the world than they wanted to see from up there! There was no question in their minds that they were going to fall off the mountain and die. We had a rebellion on our hands.

Some other personnel and I talked them down off the mountain. It was a slow, obviously terrifying, trip. They held their assigned piece of fire line that day, although they really weren't tested. Their fear of heights forced us to hike them back to camp that night. They were not happy campers. Most of them agreed to stay, providing we used them down in the trees, and didn't get them up where they could see like that again. A couple of them simply quit. When a firefighter quits, it's up to them to figure how to get home. We stretched a point and caught them a ride in supply trucks to Bozeman. They had to pay their own way back north. In retrospect, I suspect that I would experience similar fears if I suddenly found myself in one of their fishing boats in a rough sea.

Some major event had obviously knocked down a bunch of trees on a couple of ridges across the canyon, north of the fire. I could see the sun reflecting off pieces of metal over there. I asked one of the local folks what I was seeing and found out that a B-52 bomber from Great Falls had somehow underestimated how high the mountains were several years earlier. They had ricocheted off the first ridge and came to a sudden stop on the second. All personnel on board were killed.

I spent the better part of one afternoon trying to get some key line completed. The fire kept crowning and making major runs toward a ridge. I wanted to see what was up there. I established contact with a lookout we had posted across the canyon, told her where I was headed, and asked that she advise me if any major runs were coming my way.

It was a long, hot climb, frequently in burned areas inside the fire. The location was hot, and obviously not safe. I saw what I needed to and went back to camp to discuss things with the other Operation Section Chief before the planning meeting. I maintained close contact with the lookout while I was in the area, but forgot to tell her when I left.

I had been off the mountain for a few minutes when she called to advise me that a major crown fire was running upslope right towards where I had been when I last checked with her. She seemed disappointed when I advised her that I wasn't there.

"Well, just where are you?" she requested, obviously wanting to help where she could.

I had to radio back that I was lying in the shade on the deck of Peckipaugh's cabin, sucking on an ice cold pop.

There was a brief pause on the radio, than an obviously hot, tired male voice simply stated, "Kill." I never did find out who coveted my pop.

We had a couple of affirmative action problems on

the Six Mile Fire. As I've mentioned before, some women can fight fire as well as some men, and race doesn't have anything to do with any of it. On the other hand, there are some men and women who can't pop popcorn without suffering third degree burns. In this case, I think both individuals were basically good fire fighters. One just lacked sufficient experience to be in her assigned position, and the other had more pride than experience.

We had a MIF crew with a female crew boss. Her crew was made up primarily from whoever was left, since most of the regular MIF crews were already fighting fire someplace else. She had a tough job with a crew of leftovers. They made up a marginal crew at best, one that would have caused anyone problems. They showed little respect for the female Crew Boss.

We assigned them to a pretty calm portion of the fire. Their major assignment was to locate and control a couple of spot fires. They usually were sitting around whenever I happened along that section of the fire. I would get them back to work, and encouraged their Strike Team Leader to give them a boost whenever he could. The fire started picking up one afternoon. I was working with some crews in a particularly troublesome part of the fire when I heard what sounded like a jet airplane behind us. I called the Division Supervisor for that area to see what was going on. The wind had changed in the area with the questionable crew and the fire just took off. It was running east, away from anyone so there wasn't any danger. We just didn't need that much fire.

Both the Strike Team Leader and the Division Supervisor had told the whole crew they needed to do a better job a few hours before. When the Strike Team Leader and

Division Supervisor ran down to see what was going on they found the crew sitting down eating lunch and sleeping, completely unaware that the fire had taken off where they were supposed to be working. We weren't happy, and sent the crew down the road in disgrace. The crew boss filed an affirmative action complaint that cost the tax payers a bundle. She claimed the only reason they got fired was because the fire overhead team was prejudiced towards her and her Native American crew.

The other incident involved an employee who came in with some pretty top personnel we received from Yellowstone Park. She was a hard worker and did a good job as a Field Observer. We eventually worked the fire down to where we didn't need that position until we started a major burnout in a few days, designed to corral the section lost by the newly fired crew.

The lady was assigned as assistant helicopter manager in a training position. She was interested in that position, although she had no training or experience in the job. "Swede" Troedsson was running the helicopter show. There probably isn't a stronger proponent for integrating women into fire positions than Swede.

The Park lady was now assigned to do something where she no training, experience or knowledge. Logically, she should have depended on Swede and what she could pick up in the fire line notebook to bring her up to speed. She and Swede had a personality conflict when she wanted to do things her way. When time came to rotate her back to her field observer position, Swede recorded that she needed specific training and experience before she would qualify in the helicopter position, which was correct. She was used to receiving only outstanding ratings for whatever she tried and came unglued. She filed an affirmative action complaint against Swede, the champion for women on the fire line.

We worked hard getting ready to corral that portion of the fire lost by the problem crew. The least expensive and obviously safest way to corral that breakout was to conduct a major burnout that would run it into the rocks at timberline and keep it from going farther in the timber up the canyon.

We had lots of line to build, pumps to get into position, and supplies to order. We found a helicopter kit that dropped ping pong balls filled with flammable material to ignite the burn out. A chemical is injected into standard plastic ping pong balls and they are shot out of the helicopter as it flies along the area to be burned. The amount of chemical injected into the ball determines how long it is before the ball ignites.

It sounded like an exciting time to me. We had everything set up to burn one afternoon, then had to delay the burnout because the machine that fired the ping pong balls wasn't working right. I returned to camp to find new orders waiting for me. I had forgotten how long we had been there.

The boss had called to order me home for the big range review. I was headed down the road just as they got everything working. I could see a big smoke cloud in the rear view mirror as I drove away. They were conducting the big burnout without me. I wish I could have been there instead of looking forward to a field review with a bunch of specialists who couldn't agree on solutions to some serious problems.

CHAPTER 40

Harrison Creek

It looked as though I would be lucky to fit in one fire a year around normal duties. There were several calls to see if I could go. Something always interfered.

The next call that I could accept came almost a year later. A dark night in late August, 1991 found me traveling to the Harrison Creek Fire on the Lewis and Clark National Forest, between White Sulfur Springs and Great Falls, Montana. I managed to miss all of the deer spending the night on the highway along the Shields and Smith Rivers, and met the fire team at the Belt Ranger Station about four a.m.

Terry Dansforth from Yellowstone National Park was Incident Commander. Terry had been Operations Section Chief on my team a few years before. Now I was filling in as Operations Chief for his team. Since I was filling in on the team, I got the normal night line operation. We were dealing with some extremely heavy fuels in some country with a significant history of severe fire activity in recent years.

The initial briefing indicated that the fire was located between an area that had been heavily logged in the 1960's, and an area that had burned a few years before. Somehow the logged area had been added to a "roadless" inventory and was included in a wilderness proposal being pushed by some environmental groups. The large clearcuts were overgrown with a dense jungle of reproduction that had never been thinned because of the controversy over the roadless status, although there were lots of roads and cutting units involved.

The Sandpoint Fire had burned an area on the other side of the fire a few years earlier. It recovered with a dense stand of grass, forbs and other native vegetation.

The fire behavior officer predicted that things would go to pieces in a hurry if the fire reached either the dog hair lodgepole pine regeneration on the old clear cuts, or the dense grass stand on the old burn. My experience indicated otherwise. I argued that we could use either the clearcuts or the previous fire as fire breaks. We cleaned up logging debris on 1960-1985 clearcuts in lodgepole pine so well that they remain essentially fire resistant. There simply is not enough fuel on the ground to carry fire. Even on dry years, perennial grass vegetation that returns following a fire will remain essentially fire resistant for up to ten years. Fall frost can change that, but it wasn't fall yet. We didn't have to wait long for the fire to reach the clearcuts, which proved to be excellent fire breaks. The fire never reached the previous burn so I didn't get to show how that vegetation wouldn't burn.

The Forest personnel were concerned about a cabin

authorized under a special use permit that was in the fire's path. The cabin had been built illegally as a hunting camp on an old mining claim. Such abuse of the mining laws was fairly common until about 1972. Someone on the Forest apparently had "legitimized" the illegal use by issuing a special use permit-a not uncommon, if unwise, approach. It doesn't solve the illegal use, just avoids controversy. I suspect some political pressure on forest officers was involved at the time. Protecting the cabin wasn't high on my list. I had fought this illegal type of use most of my career. Quite frankly, I hoped the fire would take it out. Unfortunately, it never got that far.

We had some excellent crews on Harrison Creek. I lucked out and drew the Chief Mountain Hotshots as part of my shift. This crew comes from the Blackfeet Reservation around Browning, Montana. Their crew boss' name was Alan Still Smoking. Still Smoking is big enough to arm wrestle grizzly bears, and ran one hell of a crew. They built and burned out a lot of fire line with a minimum of supervision. All I needed to do with a crew like that was to make sure they understood what was needed, get them necessary supplies, and get out of the way.

The fire burned in some heavy fuels but the weather gave us a break with cooler temperatures and rain forecast about five nights into the fire. It was cold and cloudy when we went out that night. We started to pick up some rain shortly after dark. The rain changed to snow by one a.m.

Montana weather tends to favor extremes, so it wasn't that much of a surprise. Now I had to worry about hypothermia in the crews. I called all of the crews on their radios and got them headed back to camp as soon as it started snowing. Five nights on the fire line is tough enough. Add an extreme change in temperature, get everyone wet, and we had some pretty sad looking folks when they struggled into camp. I was pretty frozen myself when I confirmed that we had everyone off the line and was able to crawl into my own sleeping bag about three a.m.

We didn't get that much snow, but we all shifted to day crew. I worked the line while the regular Operations Section Chief handled planning. Things wound down rapidly after the storm. The most dangerous thing I saw one afternoon occurred when a family of grouse wandered through the Chief Mountain Hotshots. The air immediately filled with rocks, sticks, shovels, pulaskis, and anything else that might substitute a roasted grouse for normal fire line chow. Luckily, none of the fire crew got in the way of a hand-guided projectile. The grouse weren't so lucky.

CHAPTER 41

The Turkey Fire

A general drought continued in the west in 1992, but the Beaverhead had a lot of things going on in addition to fire. It looked as though I wouldn't be able to work a fire assignment in, although I received several calls. Summer slid by in a haze of paper work, appeals, law suits and discussions with irate special interest groups. We were well into fall and I hadn't been able to get free when a fire assignment came up.

We participated in a family Thanksgiving dinner with my mother and some other family members in Mackay, Idaho, then returned to Montana for some serious duck hunting on Friday of Thanksgiving weekend.

An unexpected fire call came about midnight. I had some pretty critical things going on the following week, but Tom Alt, the Incident Commander, was pretty desperate. Most personnel were off somewhere on Thanksgiving break. I couldn't turn him down. Tom had worked for me when I was Incident Commander. He's a good hand and really needed an Operations Section Chief. Extreme winds had whipped the Rocky Mountain Front in the Great Falls, Montana area Thanksgiving Day. These winds, recorded at over a hundred miles per hour in Great Falls, pushed a fire out of timbered areas on the Stanford Ranger District of the Lewis and Clark National Forest. It burned with amazing speed and intensity onto private lands to the east. I was off to the Turkey Fire.

Montana is well into winter by Thanksgiving, although little snow had fallen to date. I stuffed as many of my winter clothes as I could get into my fire pack and headed for Stanford, along with some others Missoula Dispatch had rounded up in Dillon.

We arrived at Stanford just before daylight. They had obviously encountered some extreme winds. There were lots of down trees and limbs scattered around. I couldn't figure out what all of the objects were that were scattered along the road as we drove into town until I realized that they were shingles. The wind had literally ripped the roofs off of several houses, one shingle at a time.

The district had a lot of slash piles, made up of debris left over from timber harvesting in the area. They had planned to burn them that fall but had been holding off waiting for more snow. Unfortunately, big game hunting season was open. Some of the hunters apparently found the slash piles attractive as warming fires. Of course the piles created huge bon-fires. The district had spent a lot of time trying to put out all of the fires that resulted. Either some of the piles weren't completely out or some hunters had lit some new ones before the high winds hit.

We didn't know what we would find when we went to

the fire line that morning. It wasn't pretty. The wind had simply picked the fire up at the point of origin, whipped it into the tree crowns and sent it screaming down on anything and everything that lay to the east. It ran out of timber near the forest boundary, but didn't slow down there.

A lot of the local ranchers had fields in the "Conservation Reserve Program", a subsidy program administered by the Soil Conservation Service (now Natural Resource Conservation Service) that pays land owners not to farm erosion prone soils. Some of these fields had dry grass standing over two feet tall in them. The fire burned well in this freeze-dried vegetation.

Unfortunately, the ranchers had just received special authorization to graze the CRP fields with their cows and still get paid the subsidy since the drought had reduced forage elsewhere. At least one rancher had just put a couple of hundred cattle in one of these fields. The cows couldn't escape the fire because of fences and the fire simply burned over the whole herd.

Some residents, including some hunters using a cow camp, told scary tales about the fire behavior that night. One of the hunters wrecked a motor home while trying to flee before the flames. He was rescued by other hunters following in a pickup truck. Thankfully, only a few marginal outbuildings were in the path. Several were burned, and immediately gained considerable value as ambulance chasing lawyers found out about the fire and contacted the affected owners. A primary power line was in the fire's path. A lot of the large wooden poles were burned off for a major part of their length. Some local fire crews reported major electric arcing between wires as the smoke created contact between wires and they burned through shutting off electricity for quite a bit of Montana.

The wind had been so strong that it simply blew the fire out when it reached some exposed ridges with low fuel. It burned well elsewhere, including across fields of grain stubble.

The one bright spot was a couple of missile silos manned by the U.S. Air Force. These silos contained inter-continental ballistic missiles with atomic capability. The fire didn't seem to bother them much. They were supposedly immune to atomic blasts so I would hope they were also immune to wild fire. The Air Force didn't invite us in to check anything.

The fire essentially burned itself out when the wind died down. However, the Forest wanted everything out. Not only was there potential for another wind storm that could carry the fire farther, but they were also very aware of the high potential for litigation. The public, especially lawyers, know that if they think they have been damaged in any way, and the government might somehow be involved, sue the government. They probably have a sure thing, since everyone knows that the government has an unlimited amount of money and must be held responsible for everything. I'm sure there were lawyers on the phone chasing potential clients as soon as they heard about the fire.

Most of the remaining smoke was in sawdust at an abandoned sawmill site, along fences where tumbleweeds had accumulated over the years, along the edges of fields where plowing had buried vegetation near the surface, and in logs and stumps in forested areas.

The weather was a major factor in managing crews. A light skiff of snow accompanied the wind on higher elevations within the fire. Temperatures hovered around zero each night, and generally didn't get above freezing during the day. Instead of a tent camp, we put the crews up in the school gym in Stanford and fed them in the school lunch room.

Caring for the engines was another matter. Water and plumbing froze quickly in this equipment if left to the mercy of the elements. We found a few heated city and county buildings and garages to keep the engines in when they weren't on the line. We probably bought up all of the anti-freeze in Montana to add to key components while the trucks worked the fire line.

We sent the Chief Mountain Hotshots and the few other crews we were able to pick up into the areas not accessible by roads. You don't often see fire crews on the fire line wearing winter pacs on their feet, mittens and down coats, but we had them on the Turkey Fire. We sent the engines and bull dozers into other areas. Nobody really froze, but it was tough to convince the crews to put out all of the fire they encountered since that constituted the only warm spots in that part of Montana.

We experienced an air inversion in the area. Cold air is heavier than warm air, so the cold air sank to fill the valleys. The wind continued pretty strong at higher elevations, especially in the air along the ridge tops and just above the fire. There was little wind at lower elevations.

I was flying in a helicopter over a section of line manned by the Chief Mountain Hotshots one afternoon. It was tough to tell smoke from blowing snow from the air. The pilot commented on the wind. His instruments indicated that we should be going sixty-five miles per hour, yet we were hardly moving at all. The wind was whipping us around quite a bit as well. I decided that if that section of line was that important, I had best check it on foot. We returned to the helispot in Stanford. The wind wasn't blowing there.

I made a point of making our bulldozers available to help bury cattle killed in the fire. None of the ranchers took us up on the offer. I suspect lawyers were advising clients that such assistance might weaken their opportunity to make big bucks off the federal government when they initiated the law suits. These animals obviously were extremely valuable and got more so every time a rancher talked to his attorney.

The area is where a famous western artist named Charlie Russell spent a lot of time and painted most of his pictures. The fire almost reached an old cabin where he stayed during the tough winter in the 1880's during which he painted a picture titled "Waiting for a Chinook" or "The Last of 10,000," depicting a cow and calf that are starving to death surrounded by wolves. It's beautiful country, even with a skiff of snow and a lot of black from a fire.

We had things pretty well out before school started Monday, so we could turn fire camp (school) back over to the kids. I also made the important meetings I had the next week.

The government paid millions in law suits on the Turkey Fire.

It rained in 1993. A few fire crews were sent east to help with flooding along the Mississippi River. Otherwise, most of us stayed home.

CHAPTER 42

Into Retirement

April 3, 1994 was my fifty-fifth birthday. I had some major career decisions to make. I could take regular retirement with thirty years of service at age fifty-five. Counting military time, I had thirty-seven years of federal service. There was a two percent increase in retirement pay for every year over thirty for something like eight years over thirty which meant there wasn't much incentive to stay. I had dedicated my life since 1957 to managing mountains for the U.S. Forest Service, including the two year interruption to help defend my country in the U.S Army. Essentially no important management was being accomplished on the ground, thanks to ongoing political and legal action.

I always wanted to retire while I was still young enough to launch an independent career. I had been working for someone else since I started working in my parent's store and slaughter house when I was eight years old. It would be nice to be my own boss. Lois and I discussed the options of trying a combination of operating a scenic/historic/environmental education based tour service and consulting on private land management. We also had some specific interest in the Lewis and Clark Expedition since my grandmother's grandfather buried Jean Baptiste Charbonneau, the baby on the Expedition, in 1866.

Yet I had entered the U.S. Forest Service as a smoke chaser. I wanted to leave the same way. I hated to leave after a fireless year following a political fire such as the Turkey Fire, where things were essentially out when we got there.

The only incentive to stay involved a law suit filed by the National Wildlife Federation objecting to allowing livestock use on the Beaverhead-Deerlodge National Forest. At first I considered it a personal affront since I knew our program was possibly the best in the National Forest system and we were working to improve it. Their attorney advised me they knew that our program was among the best. If they won against us, they could win against units with inferior programs.

I was looking forward to the challenges. I wanted to continue livestock grazing on rangelands where the vegetation evolved with grazing use. Granted the evolution originally involved bison and other wildlife. However, bison are gone and aren't coming back. Private land now occupies a significant amount of historic wildlife winter range, and available winter range restricts wildlife populations in Montana. Managed livestock grazing remains a key tool to help meet vegetative management objectives.

Then the whole retirement picture changed suddenly. A new Congress was madly rushing into a plan to drastically reduce the number of government employees. It didn't matter where the reductions came from and what the consequences were. I agree with the need to cut taxpayer costs, but it isn't something to be tackled blindly. There are too many valuable natural resources and other needs that suffer from hasty decisions.

Since I was more than fully qualified to retire, Lois and I were given three days to decide whether to participate in a "buy out" program or to choose normal retirement later with no incentive. I had to retire by May 3, 1994 for the incentive. I accepted. A lot of great people with a world of valuable experience retired in early May, 1994. The U.S. Forest Service lost a lot of experience and expertise that spring.

The serious drought conditions continued in the western United States. Fires started early, with some of the most serious in Colorado. A lot of people have built expensive homes in historic fire vegetative types along the east slope of the Rockies. We'd successfully kept fire out of this area for all too long. Nature was catching up.

It didn't take long to see that the fire control system lacked the depth to cover needs on a serious fire year, especially since a lot of key personnel had just retired. Several of us tried to get back on to help out. Few, if any, of those in personnel management know anything about fire needs. They maintained it wasn't legal for us to go back to work for the Forest Service after retiring.

Than a group of firefighters were burned over and killed in Colorado. Details identified that all of the victims were very experienced smoke jumpers and hot shots who were simply taking too many chances.

They simply should not have been where they were. They had too much training and experience not to know the chances they took. The fellowship of fire unquestionably contributed. They knew they could depend on each other to do almost impossible things because they had done it before. They got caught on the Storm King Fire in Colorado in 1994 by making many of the same mistakes that killed fire fighters in Mann Gulch in 1949.

One outcome was that the need for more experienced overhead was identified. Forest Service personnel managers still dictated that recent retirees couldn't go back on duty for the Forest Service. However, they couldn't stop us from signing on as employees for state and other firefighting organizations. I signed up to fight fire for the State of Montana.

CHAPTER 43

Klickitat Complex

I didn't have long to wait with all of the fires that were burning. Bob Michael's Type One Team was headed for the Yakama Indian Reservation in Washington and needed a Division Supervisor. I was going back on the fire line again!

We took over the Klickitat Complex. There were eight fires involved in our complex. Some were in very heavy fuels on rough terrain with lots of potential. We were able to size up some of the fires as we flew in to the reservation.

Michaels caught me after the initial review and said "Having you on the team gives me a lot of flexibility. With your back ground, I can work you in wherever I need the help. You're now night Operations Section Chief."

My opinion of night shift has never changed, but what could I say? The team was using the concept of both line and Planning Operations Section Chief that we had worked out earlier. That leaves a hole when it comes to close attention on the night shift. We obviously had the opportunity to accomplish a lot at night on Klickitat, and close supervision was needed.

Base camp was at an old Forest Service ranger station that had been transferred to the Yakama Tribe, along with a lot of land north of Mount Adams, when Nixon was President. The country is spectacular, with a lot of excellent timber. The tribe had selectively logged a lot of the area around where the fires were burning. We had good access on logging roads to much of the area. The primary exception involved three fires burning in the river breaks east of the Klickitat River.

I remain impressed with some of the land management the tribe has initiated. They have hunting seasons for tribal members and consequently have a lot of game, something that is lacking on a lot of reservations. Their timber practices extracted more valuable tree species such as ponderosa pine and western red cedar while leaving less valuable trees such as grand fir which favors some wildlife such as spotted owls.

We were able to use bulldozers backed by engines quite successfully on a lot of the fires. One fire was high on an unroaded mountain and didn't seem to have a lot of potential. We set up a spike camp on that fire, sent in a couple of crews, and supported them by helicopter.

We made great headway on night shift, either completing line in areas too hot to tackle during the day, or burning out lines that had been put in by the day shift. It was a lot safer to burn them out during the early part of the night rather than trying in the heat and windy conditions during the day.

Those fires could really move. We were just changing

shifts one evening when a major blowup occurred. The day shift was supposed to be off the line and night crews were just reaching the line. When the run occurred, I was across the Klickitat River, west of that fire. We were showered with all sorts of burned branches and bark that had been sucked up in the convection column. Everything we could find was cool when it hit the ground, but there had to be some hot stuff out there somewhere. We found a lot of new spots that night, including some that had crossed the river. We held the fire at the river.

A crisis developed along the line and I was in no position to help. We were using a lot of contract crews. Contract crews have improved significantly since my problems with them on Corn Creek in 1961. They normally have good, experienced overhead from crew boss down. They work long, tough, dirty hours, so it is difficult to get dependable line workers, especially from the white welfare unemployed ranks. Most of the members on this crew were Hispanic. Only a few could speak English. I suspect a lot of them were on work permits out of Mexico. A squad from one of these crews was intent on completing a section of the line before pulling off the day shift. They somehow missed the call that the fire was making a major run. It cut them off and they were in trouble. They were justifiably scared when they called on the radio. Fortunately, Barry Hicks, a very experienced Air Operations Officer, was flying the fire at that time. Barry located them from the air and directed them to safety. It was close, and they had a pretty good scare.

The blow up occurred between two of the fires giving us problems in the river break country. It helped us decide to burn out the area between the fires and reduce the amount of line we had to build and hold in that tough terrain.

Both the night and day operations continued to make progress. I still had problems trying to get some sleep. We found a shady spot where we could try to sleep during the day but there was still all of the noise from camp. It's hard to get much sleep with generators, trucks, bulldozers, helicopters and people running around, not to mention the heat, insects and other distractions.

Golden mantle squirrels presented an interesting pest at this camp. The entire area was overrun by them and they were very brazen. They begged food and got into everything.

I woke up about eleven one morning to realize that I wasn't alone in my small pop up tent. Before I could move, the squirrel went crazy. He dashed madly about, squealing loudly. I sat up and thrashed at him with a boot which probably didn't help. He finally escaped through an unzipped spot on the door and ran, still squealing madly, into a pile of rocks.

I didn't have any food in the tent so I couldn't see what caused his excitement. I tried to get some more sleep with limited success. I finally gave up and headed for the shower. I discovered what had upset the squirrel. He had been rummaging around in my overnight kit. It hurts to admit just how many miles of fire line I have behind me, not to mention the years that keep flying by, but that isn't all that hurts. When a person gets that much country behind him and still likes to roam the mountains, sore muscles come with the territory. I had some "Ben-Gay" analgesic balm to help heat up aching joints. The squirrel chewed a hole through the tube and apparently got a mouthful of some pretty hot stuff. Everything in the kit was covered with Ben-Gay.

Secretary of the Interior, Bruce Babbot, made a political trip to the fire line. There were lots of communities and very high value improvements at risk on other fires. These fires had to have priority for manpower and equipment over more remote fires, such as ours. We were burning up some pretty valuable timber but no communities were threatened. However, the tribal council decided we weren't getting the attention we deserved. Babbot was on a tour of fires around the west, politically supporting the firefighting efforts. He's an old smoke chaser, and deserves credit for knowing a lot more about wildfires than most politicians in Washington D.C. He spent a couple of hours on our line which gained quite a bit of press coverage. He spent a lot more time trying to assure tribal leaders that all possible support was being committed.

I had picked up a trainee in the Operation Section Chief position by then. He was a good hand out of Arizona and it didn't take long to get him qualified.

Meanwhile, the Yakima Tribal Council requested help from our overhead team in training a hundred and sixty of their people as firefighters. The team had assigned the safety officer to assist with that objective. He picked up some state personnel to help with the classroom portion. Eighty of the new firefighters were completing classroom work the next day.

Michaels asked if I was ready to turn the night operations position over to the trainee and take over the job of taking the new crews on the fire line for their first day to conclude their training. The new challenge sounded great. It was also a day assignment that promised a vastly improved opportunity to get some much needed sleep.

We were to meet the new crews at Tribal Headquarters in

Toppenish, Washington at eight that morning. I would take the first eighty on the fire line on the Lakebeds Complex several miles east of our camp, while the second eighty started their classroom training. Lakebeds had good line around one of their fires, and basically needed crews to patrol the line and mop up any fire they encountered. It was a good place for a green crew to start. We headed for Toppenish.

The lawn in front of tribal headquarters was a mass of confusion when we arrived. It was full of new firefighters plus friends and relatives who came to see them off. We got the fire fighters more or less separated from relatives and the curious and issued them their standard fire packs, standard supplies and fire retardant Nomex clothing.

It took a couple of hours to get everything sorted out. We had all sorts of situations develop. Most of the aspiring fire fighters were young with several women on each crew. Several questions were raised by the ladies concerning what they had viewed as a promise by the trainers that they would be treated as equals by the males. Some of the ladies apparently felt they weren't treated equally at home, so why should that change on the fire line? Try to answer that one!

A lady who was probably about my age beat most of the others in getting her stuff together. She caught me as I walked through the crew members, checking their progress. I'm no beauty, but her face reflected the disappointment and hardships that go with over half a century of reservation living.

"I want to help so badly," she said, "but I have trouble hearing. I might not hear someone holler 'get out' if things go wrong. I don't worry about me, but what if someone gets hurt trying to rescue me?"

That grandmother had all of the care and sincerity in the world in her heart. It reflected openly on her wrinkled face. And she wanted to work so badly. I couldn't tell her that maybe she shouldn't go.

"Can you find a friend to work beside?" I asked. "Let them know you need help hearing, and they can let you know if anything important is said."

"Ah, yes," she said, her problem solved. "I'll work beside my granddaughter," she said, pointing her out in the crowd.

Granddaughter was having problems of her own. She was a very good looking young lady. The "web gear" that supports a firefighter's pack, canteens, fire shelter and other supplies on their back and waist while they're on the fire line is obviously designed with men in mind. Some of the major straps in front cross at breast level, something that obviously can cause discomfort for ladies with large breasts.

I was standing there trying to figure out how to gracefully assist with that problem without embarrassing everyone, especially me, when a lady came bounding over from another crew.

"Hell, honey, I had the same problem! Here, tighten this, lengthen that," she said, adjusting the many straps that hold everything together. She was immediately surrounded by women from other crews. There is a God!

The next crisis involved three huge young men. They looked more than qualified to play defensive tackle on any professional football team. Actually they were big enough that any one of them probably could have made up the whole line. I thought that fire resistant Nomex clothing was made big enough to fit a bulldozer. I was wrong. We didn't have pants big enough to even come close to the two size 44 and one size 46 waists involved. The same applied to the XXXL shirts required.

They were so sincere. They had spent three days getting trained up to this point, and they were really keyed up to go fight fire. Yet we couldn't put them on the fire line without the proper safety gear. Someone suggested I take them to base camp and see if we could find anything bigger there. I doubted it, but didn't see a choice.

A tribal council member caught me before I could pursue the clothing problem farther.

"I have come to give a Song and a Prayer when it is time for our people to go," he said. I figured I had lived on a song and a prayer for most of my life. I wasn't sure what he had in mind though and asked for clarification.

The Yakima practice a lot of tribal tradition and are a very religious people. Before they send their people off to war or equivalent, a tribal leader performs a ritual designed to help those involved through the troubled times they may encounter. It sounded like a great idea. I assured him that I would call him forward just before we boarded the buses.

I was still concerned about the three big guys. I gathered up six of our biggest pairs of pants and shirts and headed for the tribal personnel office. I found a lady there and explained my problem. She had a sister who was a hair dresser who had a friend who was a seamstress, or knew a seamstress, or something like that. She thought they could fix things up if I got authorization to pay for the alterations.

I headed for accounting. A lady there assured me that she had a sister who had a husband who had a friend who could fill out the necessary forms so it should be okay, if

I wrote down what was going on and signed it. I told her I would pay out of my own pocket if necessary. I figured I was close to getting that detail settled and headed back to the crews. The crews were ready to go.

I headed for the sidewalk between the firefighters and their buses. I have a pretty good idea how my drill sergeant felt when he lined a bunch of us draftees out in Army basic training. But these were all volunteers, and they were ready.

The tribal leader came forward. He told me that everyone needed to take their hat off during the ceremony, but didn't want to embarrass his young people since they should know. I directed that request over the loud speaker to those of us non-tribal members who were present, which served the same purpose. Everyone's hat came off, regardless of race. I turned it over to him. He pulled out a small hand bell. He explained the ceremony and that he would run through it in Yakima, and again in English.

That was an impressive ceremony! He proceeded to accompany his chanting song with the bell, and followed with the prayer. I understand no Yakima, but the ceremony made me wish that I did. In English, he assured us that God would watch over us and help us conquer the fires. The animals and plants, and the earth and water knew that we were going there, and that we were there to help them. We might not see the spirits of those sacred things but we needed to know they were there. They knew we were doing our best to help them. And they appreciated us for doing so.

Then it was over. All I could say was "Load 'em up", when he finished. We were on our way.

It was about fifty long, steep miles to the Lakebeds Fire

Camp. I was riding on the first school bus in a four bus convoy with one of crews. The buses had excellent Citizen Band radio contact. We hadn't gone far when the lady driving the bus in the rear of the convoy started to complain that her engine was overheating. She started to drop way back. We were approaching some key road junctions. I was getting concerned about how long it was taking us and more concerned about her getting lost.

I discussed what might be the problem with the lady driving our bus. She assured me that the buses were in good condition and there shouldn't be any problem. They were designed to carry up to forty-eight passengers. Climbing the hill with the twenty adults plus their gear wasn't a load. The lady behind was still complaining and said she was going to have to stop.

Our driver knew a whole lot more about vehicles than I ever will. She had just started explaining the automatic transmission to me when she grabbed her CB mike and called the rear bus.

"Hey, Betty," she called. "What gear you got that baby in?"

"Drive," was the reply.

"Jam that sucker in low," our driver stated. "I've been climbing all the way in either second or low. You try climbing this mountain in drive and the transmission keeps shifting back and forth and overheats. I thought everybody knew that."

I didn't and obviously Betty didn't, and the chubby man driving one of the other buses later admitted his bus was heating up as well until he heard our driver and shifted down.

It didn't take a couple of minutes for Betty to call back

and admit that her engine was cooling down. Our driver simply said "Great, now quit 'cher bitchin' and let's get these fire fighters where they belong."

We checked in with logistics as soon as we reached base camp. They showed the crews to their sleeping area. While that was happening, I checked supply for extra large clothing. They didn't have any. I found the three big guys, and they were assigned to a camp crew. Their disappointment was obvious. They signed on to fight fire. Now they were dumping garbage cans and policing the camp area.

It was after noon before we finally reached the fire we were assigned. It was on fairly easy ground, which was good. It had made a major run ahead of favorable winds, slowed when the winds died and had been lined. Now it just needed patrol and mop up. We lined the crews out and started to work. Each crew had an experienced crew boss from the agency, and we let the Crew Bosses do most of the training. It went very well.

We had been assigned Strike Team Leaders, and everything seemed to be very much under control. All I really needed to do was check back the next day to see if there were any problems. It was an interesting afternoon, then back to camp. I crawled into my tent at Klickitat Base Camp about midnight. I had to be back in Toppenish at 8 a.m.; so much for catching up on sleep.

I took in a couple of hours of the classroom training for the second group of eighty the next morning. It helped me understand how that section was put together, and how it might affect their on-the-line performance. Then I checked in with the personnel specialist who had a relative who had a friend who knew how to sew.

They were all there. So were the BIG clothes I had requested. They had done an excellent job. I signed for everything, even admitting that I would pay the almost fifty dollar bill personally if I had to. I'm not sure how that was resolved, but I never got the bill.

I drove to the Lakebeds Base Camp to check on how the crews were doing. The three big boys were still in camp. They really didn't have that much to do and were getting discouraged. I told them we were going to change all of that. They were going back to their crews. I pulled out the tailored clothes. I don't recommend doing that. Being hugged by three huge Indian men at one time isn't conducive to maintaining bone integrity. One of them actually cried. "Why did you do this for us?" he asked. "No one has ever cared that much before." It felt good. I don't know how it worked for them in the long run. At least they had a shot at doing meaningful work. Maybe it helped.

My next step involved getting the next four crews on the line on the Klickitat Complex. Basically, it was a repeat of the old routine. Again, the "Song and a Prayer" ceremony was most impressive. I wish we could have gotten it on a video tape. We were making pretty good progress on both complexes, although high winds could have changed that fast. Conditions still hadn't allowed us to complete the burn out between the two fires in the breaks.

We got the new crews integrated into the fire line organization and I was out of a job. We were now well into the burn out between the two fires, although some of the line along the top of the breaks remained to be completed. I was assigned as Division Supervisor to help complete needed line, burn out, and hold one of the divisions involved.

The fire made several runs at our line. It even jumped the line in a couple of places, but we picked them up. We had a pretty rugged piece of country, but an elaborate water system was set up to help. We pumped water uphill from a stream with a portable pump, then picked the water up with a series of engines to pump it on up the line. We put out a lot of fire with water.

I had some personal concerns at this time. Our oldest daughter, Jan, was expecting our first grandchild around August 1. Both she and her husband, Chuck, had to turn down fire assignments for the big event. And here I was a couple of states away with very limited communications. We had one cell phone in camp, but had to drive about a mile out of camp to get enough towers to make a call. I woke Lois up more than once when I was late getting back to camp. The word was always the same. I wasn't a grandpa yet.

I came dragging back to camp about eight p.m. after a particularly trying day on August 6. I had word waiting that our youngest daughter, Teri, had called from the consolidated dispatch in Toppenish. I phoned her there. Teri worked for the Curlew Job Corps Center on the Colville National Forest in Washington. She had been out on other forest fires managing camp crews from the center for three weeks and finally had a day off. She decided old Dad needed company for supper and headed for the Klickitat Complex. We had both been keeping in pretty close contact with home, so at least she knew where I was. She got as far as Toppenish. Visitor traffic isn't allowed on fires. Basically, the personnel in dispatch understood but they couldn't let her through because of the precedent it would set. Tribal members all had family and friends on the fire line that they would like to visit too. I told her to hold tight. I'd grab a truck and meet her at dispatch in forty-five minutes.

She also said she had called home, and Mom wasn't there. I suggested that she try to track Jan down, since that might be where Lois was. My body kept telling me all it really wanted was something resembling food and a sleeping bag. On the other hand, I hadn't had supper with Teri since my retirement party that spring. A truck was heading for town to get supplies. Indians understand family a lot better than most folks. The driver agreed to wait in case supper took longer than loading his supplies. We headed for town.

Teri was talking on a phone in a corner when I walked into dispatch. She was jumping up and down, obviously quite excited. She handed me the phone when I walked over. She was talking to Lois at the hospital in Sheridan, Montana. Tessa Rose Bowey was thirteen minutes old. I was a grandpa! I don't remember what "Aunt" Teri and I had for supper. I remember great conversation and the excitement of knowing we were somehow part of a new generation. The excitement was contagious. The young waitress brought us a free slice of pie each to help celebrate. She refused payment, but got a nice tip. Then it was time to go. Teri headed north, promising to stop and sleep in her car if she got too tired. She carries a pistol in the car. She knows how to handle herself, so unwelcome visitors wouldn't fair well.

The driver was waiting when Teri dropped me off at dispatch. It was after midnight when I crawled into my tent. Four-thirty a.m. came awfully early. Teri left a plastic bag full of "snickerdoodle" cookies she had made special for her father. I handed them out to other overhead team members with the announcement that I had become a grandpa since we last met. There is nothing greater than loving and being loved.

We had things wrapped up, at least as far as a Type One Overhead Team was concerned, on Klickitat within another week. We turned what remained over to a Type Two Team on August 12 and debriefed through tribal headquarters at Toppenish, then were bused to Yakima for the night and a flight out the next morning.

Michaels was very pleased with the team's performance and requested that we stay on call for the next dispatch. It was obvious that the next call wasn't that far off. We could see smoke columns from lots of fires from the plane as we flew across Washington, northern Idaho and into Northwestern Montana.

It was a great team and I'd like to have stuck with them. I knew who I could depend on to do what, and the fellowship definitely was there. I did have an additional social event that seemed to outweigh that commitment. Our son, Jay, had established other priorities. He was a very lonely range conservationist for the Forest Service in Ely, Nevada until he met an equally lonely Ely High School English teacher, Susan Halferty, from Lancaster, Wisconsin. They planned to be married on August 20 in Wisconsin.

I figured that if Jay could forgo fire assignments to get married than I probably should attend their wedding too. Actually, Jay's boss was one of the "anti-fire" rangers, so Jay wasn't getting to go on fire assignments anyway. Besides, Lois advised me that it would be a very good idea that I go. After all, it isn't every year a person retires, becomes a grandpa, and gains a beautiful new daughter-in-law.

Besides, it looked like there would be plenty of fires around after the wedding.

I got home in time to see my new granddaughter, take a shower, change clothes and catch the plane for Wisconsin.

The wedding was a fun break. Wisconsin was getting most of the rain the West should have gotten, plus a lot more. The rain let up enough for a beautiful outdoor reception. Susan had to report to teaching duties in Ely on August 25 so they planned to spend most of their honeymoon driving west. Lois suggested that they take our plane tickets and fly to Montana for a real honeymoon. Lois and I drove Susan's car to Montana along the historic Lewis and Clark route and met Jay and Susan in Montana.

CHAPTER 44

Libby Complex

The next fire call came the night we got home. I was off to the Libby Complex near Libby, Montana. A Type One Team out of Washington and Oregon had the complex. It always takes a bit to determine who to depend on to do what when entering a new team, but it didn't take long for the fellowship to clarify all of those concerns. Fires surrounded Libby, and there were several other complexes in the general vicinity. Two of our fires were burning in remote country where we didn't try night activities. The rest of the terrain was steep and heavily timbered, but logging had established better access than most fires I have been on.

I checked in with planning as soon as I arrived. I had been dispatched as Division Supervisor, but they had to decide where they really needed me. While the Incident Commander and Operations Section Chief were reviewing my qualifications I discussed the fires with the planning unit.

"So you're from Dillon," one of the planners stated. "Do you know a Chuck Bodrey, or Boadry, or something like that. He came from there."

I had to admit that I wasn't familiar with that name. I asked what position he was assigned to see if that might help.

"Well, he's one hell of a hand," he stated. "He checked in as Field Observer. He helped us set up the Situation Unit in planning here. He's out on Hanging Flower, one of our tougher fires now, walking the line to gather information for the Situation Unit."

Our son-in-law, Chuck Bowey, qualifies as Field Observer but was from Sheridan, Montana on the Beaverhead National Forest. His papers probably showed he shipped out of the forest supervisor office in Dillon. I suggested that it might be Bowey, and they readily agreed. They were trying to straighten out some paper work on him, but they got so busy analyzing the information he brought off the line whenever he came back that they forgot to ask him. I didn't even know that he had been dispatched, what with a new three week old daughter and all.

The Incident Commander and Operations Chief came in then. They asked if I had any problems handling line operations on several fires in the complex.

I said that was fine with me, I'd do anything to avoid night shift. They assured me that wasn't part of the deal. What they really needed was additional overhead on night shift. Damn, I hate night shift, but I was there. It really wasn't a surprise. The fires in the complex had been going for some time. A good camp was in place, although it was tough to come by fresh, dependable crews, and a lot of critical supplies.

Northwest Montana is big timber country. Everything was bone dry. Water was critical if we hoped to catch the fires due to heavy layers of decomposing organic material under the trees.

Night shift had mostly been patrolling the line established by the day shift up to that point. We started doing some aggressive line construction and burn out at night. There were limits on where we could work. The snag problem and steep terrain added to the lack of water. It's just not safe to work crews where there is a lot of dead timber since the dead trees keep burning off at the roots and falling down. It's especially dangerous at night when we really can't locate problem trees. Add terrain so steep that we had a tough time just standing up, and we had major safety problems.

We picked up some people who knew a lot more about water systems then I do, and set out to get water onto as much of the line as we could. We had one area where air borne burning material kicked up by the winds had started spot fires way below the main fire and road that served as a fire line. The spots were really heating up in heavy fuels and steep terrain. We'd have major problems if any of these fires came roaring uphill and closed the road.

It was viewed as a trap for day crews, because of the intensity that could develop during the hot part of the day. They were able to get most of the problem snags cut, but asked if we could try to catch the spots at night when the intensity problem was lower because of higher humidity.

I flagged a route in with engineering ribbon, explained what was needed to the Division Supervisor, and sent crews in with him. I drove on to another problem area to see how things were holding on a tough piece of line. I just got there when the radio announced that a crew member en route to the spot fires had sprained her ankle. They were in heavy brush and timber and could hear the fire crowning trees below them. They were afraid to advance with an injured person.

It was a good call on the Crew Boss's part. I was sure they weren't in that much trouble, but he didn't know the country and it was dark. The crew with me had an emergency medical technician. We headed for the crew with the problem.

We weren't all that sure how we were going to get the injured person out. We had to walk across very steep terrain, with lots of ups and downs to get around logs and rocks, all through dense underbrush. A stretcher would be useless, especially in the dark. We couldn't get close with a helicopter even if we waited for daylight.

The injury didn't look that severe, although it's hard to tell about sprains. The victim was a lady with a pretty poor pair of boots. These types of injuries aren't all that uncommon, especially when there are too many fires going and people are added to crews who have poor equipment and aren't in all that good of physical condition. I'll say this for our victim in this case. She was one tough individual, and she wanted the hell out of where she was. The crew was willing to move on once we assumed responsibility for her evacuation. They proceeded to the fire where the Strike Team Leader could direct their work. We braced the ankle as best we could, carried the injured lady's tools and pack, and helped her through the toughest stuff. She hobbled through the rest and met the first aid personnel from base camp at the road about midnight.

I walked back to the spot fires that were the major

concern. The crews were getting pretty good lines around them, but the line wouldn't hold unless we could get water there. I contacted the individual responsible for that fire on the radio. He agreed to meet me on the main road above the fire if I could walk straight up the hill to the road.

It was very steep but otherwise not that bad. I made it from the top of the spots to the road in about ten minutes, straight uphill. We set up a couple of portable water tanks on a wide spot on the road, got some water tenders (large tank trucks) to keep them full, and strung hose downhill to the spots. We had a major siphon going within a matter of hours. It saved the night for that problem.

The fire rolled farther down the mountain on day shift a couple of days later. That added too much pressure to the hoses. We had to add several valves to the line to reduce pressure so we didn't blow up the hoses on the lower end before we got through.

I had a tough time keeping track of where the day shift constructed line. It's just tough to keep track of that much activity. Not only was the terrain very steep with heavy brush and timber, but smoke and darkness made things tough to follow. There was quite a bit of indirect line, and we needed to burn it out. I spent quite a bit of time ahead of the crews, trying to locate where it was productive and safe for them to work.

I was assigned a local driver who had contracted his Ford Bronco to the fire effort. He had obviously spent a lot of time "road hunting" (driving up and down roads waiting for a legal animal to step out onto a road so he could shoot it) in the area, and knew essentially every road and trail out there. He got me where I needed to go.

My driver warned me about grizzly bears before I walked

out a fire line into a big huckleberry patch about two a.m. one night. The huckleberry season was pretty well over, but this patch had obviously been a favorite for bears. Some of the piles of droppings were impressive. I respect black bears, although I've only had one charge years before. Grizzly bears are another story.

Somehow it got darker and more threatening the farther I walked out the line. And I was all alone, just me and a little head lamp that seemed to get dimmer by the minute. I started whistling to keep myself company and let the grizzly bears know I was coming. Then something that sounded really big ran through the brush next to me. Looking back, I'm sure it was a deer or elk, or maybe a rabbit, but then I knew it was a grizzly that was looking for a midnight snack. It's amazing how fast a person can get his back pack off when he is really scared. I always pack a couple of backfire fusees in case I need to help with a burn out, or start an escape fire where I can set up a fire shelter. I had a fusee in my hand within seconds. That bear might get me but I'd give him heartburn on the way down, assuming I could get the fusee lit while he was saying grace.

No bear showed up in my head lamp, so I started nervously on my way. I did a lot of whistling that night. I didn't set the fusee down until I got back to the truck. It's amazing how badly you can spook yourself out there all alone in the dark.

One of the water tenders we used to fill up the portable tanks was from Wyoming. They had to drive about a mile up a logging road to a creek where we had installed a portable pump to fill their truck. It was close to the spot where I spooked myself about bears. They were a good crew, and did a good job of keeping the tanks full. They were making fun of one of the crew members when they came back with a load of water one night.

She apparently had to go to the bathroom while they were filling their tank, so she slipped off into the brush and dark to take care of business. Then what she reported was a "really big bear" jumped up in the brush and took off. She took off the other way. She apparently did see her bear. She reported that her bear said "woof, woof", which is more than my rabbit or whatever it was did. The rest of the crew kept asking if she went to the bathroom before or after she met the bear. Her pants were so wet from working water that she wasn't in a defendable position, so she took a lot of good natured ribbing. They couldn't talk her into getting down off the truck when they were filling it after that. I don't blame her. She didn't sign on to be bear bait.

We had another scare in that vicinity one night. My driver and I were traveling between the portable tanks and the water source when we met a pickup driving too fast with bright lights on a blind turn. My driver made it to the edge of the road just in time to let the speeding truck by. I radioed back to the tanks to let the folks there know that a truck was coming much too fast so they needed to get out of the way.

An engine from Columbus, Montana was filling at the stream when we got there. One of the crew members was a Montana highway patrolman who was using some of his vacation time to fight fire. The rest of the crew had stayed to work on the fire while he went for more water. He had started the pump and went into the brush to go to the bathroom. He noticed a pickup truck stopped by the pump when he returned. The occupants were two teenage boys wearing shorts and tennis shoes.

The patrolman demanded to know what they were doing there. They claimed to be lost, jumped in their truck and speed away. This was the truck we met. We started adding things together. The fire line is a bonanza for thieves. Firefighters get very busy, and a lot of equipment is left unattended between shifts, or when not needed for a period of time. Security is one of the primary reasons for closing areas around fires. We alerted law enforcement but the kids got back on the main roads before they could catch them. We have no idea what they got away with.

We got things pretty secure on that fire, and turned our primary attention to another nearby. The area had been heavily logged in the 1960's. The fire spotted through the brush on the old timber harvesting clear cuts, and burned quite violently in the heavy uncut timber between them. We had good access on old logging roads, and used a lot of them for fire breaks. The un-logged areas were full of hazardous snags. The fire burned very hot in the heavy fuels, even at night.

My regular driver normally worked for oil exploration seismic crews around the world. He was finalizing arrangements for a new job and had to take a night off. I was assigned a lady driver with a newer vehicle. She had been driving on the fires for some time and knew the roads well too.

The day shift hadn't been able to do much work in an especially troublesome area. They asked if we could do anything there at night. I went to scout things out before dark.

The fire was burning very hot and there were lots of snags. It didn't look encouraging. We drove past a spot fire above the road before encountering a large snag that had fallen across the road. Before we could back up we heard a big "crack", and another snag fell across the road right behind us. The fire ahead was crowning, and the spot behind us was picking up. We were not in a good situation.

Fortunately, the snag behind us was a cedar that had been dead for a long time and was very brittle. It cracked in numerous places when it fell. We were able to break it up with a few nudges here and there with the truck bumper. I was able to throw enough of the pieces over the edge for us to escape. I didn't work crews there that night.

We tried again before dark the next evening. The biggest problem was that the access roads switched back and forth up a steep slope. The fire ran directly up the slope, and hadn't slowed down when it crossed the road in the heavy, dry fuels between the old clearcuts. We had to drive through the fire several times to get high enough on the mountain to reach the part of the fire that was still advancing. A day crew had completed limited work in the area that morning.

We were able to get through the problem area from the night before, but things still didn't look that good closer to the head of the fire. I left the truck and walked into heavy smoke and fire and located an area that needed to be burned out. Snags were falling all around again. I pushed things until it looked too hazardous. I was about to back out when I noticed a big black raven sitting on what appeared to be a dead person in all that smoke and fire. I walked to within ten feet of the raven before he squawked and flew off through the smoke. The "body" turned out to be a backpack hand pump. The raven probably smelled the water inside and was just looking for a drink. I'm not big on omens, but the raven didn't help. This was the thirty-seventh year that I had walked the fire line. I had never lost a person to more than minor injury up to that point. The resources on this mountain weren't worth taking the obvious risks if I pushed too hard.

We had plenty to keep us busy lower on the fire so I backed off, but not by much. We were back the next night. The day shift had cut most of the hazardous snags, we got the rest before dark, and our burnout went like a dream.

I heard an interesting discussion between Chuck, my son-in-law, and another person on the radio that night. I usually got to see Chuck for breakfast each morning. Actually it was my supper. I tried to get back to camp in time to attend the five a.m. briefing for the day shift. I updated everyone on progress made during the night and answered questions. Then I would get with Chuck and we would eat his breakfast and my dinner together before he went on day shift. I still had to make arrangements for anything I knew I was going to need that night. It was always a race to see if I could get in my sleeping bag before ten a.m. so I could get some sleep before it got too hot.

Chuck worked day shift scouting the Hanging Flower Fire in the drainage that served as the municipal water source for Libby, Montana. Primary morning access to the fire was by helicopter. In the evening, they hiked straight down the steep hill to reach a road where buses met them. Smoke tended to accumulate in the area during most nights. They had to wait for morning breezes to clear out some of the smoke so they could get in. Because of the steep, rocky terrain on the fire, they were using hot shot crews. They were supported by heavy "Sikorsky" helicopters using water buckets filled in the lake supplying Libby's water.

The Crew Boss for one of the hot shot crews claimed he had been a Navy Seal and appeared to think he was quite a mountain man. He apparently enjoyed putting anyone down who wasn't on a hot shot crew, and that sort of got to Chuck. Chuck is good in the mountains and wasn't about to let anyone out do him.

A particular area had been giving them problems so Chuck and the Crew Boss went to check it out late in the day. The Crew Boss told the crew to hike down to the buses when it was time to go. Chuck and the Crew Boss climbed up to the trouble spot and got the information they needed. Then they headed down towards the pickup point. The Crew Boss was sure he knew a short cut. Chuck told him he was wrong, but he didn't listen. It was pretty heavy going in thick brush and timber, and it was getting dark. They got separated. Chuck reached the buses and found the crew waiting for them. I could hear Chuck discussing directions with the crew boss over the radio. The Crew Boss was totally lost and Chuck wasn't having much luck talking him in. He eventually reached a ranch house a couple of miles from where he was supposed to be well after dark. They had a hard time finding the right road so they could rescue him. Chuck was feeling pretty smug the next morning.

Chuck missed seeing a major event that had things buzzing when I got up one afternoon. Those large helicopters can haul a pretty impressive load of water. The only problem they have is that the water buckets are suspended on a cable below the helicopter. Accuracy of their drop is affected by how much the bucket is swinging, the wind, rotor wash, etc. at the time the drop is made.

Making any type of aerial drops over a forest fire is pretty tiring, tedious work. There are some strict rules on how much time a pilot and ship can fly before taking a break. A company was taking advantage of one of these "down times" to install a belly tank that filled using a siphon on one of the Sikorsky helicopters.

They got the tank in place, flew out over Libby's water supply, dropped the siphon hose, and started the pump. Adjusting a helicopter in flight to compensate for taking on a sudden change in weight takes a lot of skill. I assume the pilot made the necessary compensations. Unfortunately, the pump did not shut down after the eight seconds it took to fill the tank. It simply kept going and sucked the helicopter and the three people on board right down into the lake.

We now had three people, a helicopter, and lots of Jet-A Fuel in the domestic water supply. No one was happy with the situation, especially the three soggy folks who made it to shore, plus the helicopter insurance company. A HAZMAT (hazardous materials) team was ordered. They arrived with supplies needed to contain the petroleum spill before it got into the water system. I assume they eventually got the helicopter out as well.

We were making a lot of headway on containing the fires. Stopping the major spread on these fires was one thing; putting them out was another. There was just too much fire and fuels, and the weather wasn't cooperating. It was obvious that we'd have fires going in that part of the world until they went out under winter snow.

Hunting seasons were scheduled to open, so we were getting more people in the forest although the governor postponed the seasons. The general area around a major fire action is closed to public use for security, safety and related reasons.

Somewhere in northwestern Montana that fall some extremely intelligent anti-government members of a militia group apparently decided it was unconstitutional or something to keep them out of the forest. They chose to sneak around a road block regulating traffic to a forest fire to do some hunting.

And then they stumbled onto a fire camp. Now here was a major encampment with all sorts of people running around all dressed in green pants and yellow shirts (a uniform of sorts). They were wearing helmets (hard hats). And, horror of all horrors, there were all sorts of evil looking helicopters (obviously "black" helicopters) flying around hauling loads of all descriptions. We were being supported logistically by the Montana National Guard, so there were military vehicles and helicopters involved.

The imagination is an amazing thing, especially if a person isn't all that bright to start with and has honed his skills with a six-pack or two of beer. I'm not sure where the tanks came from, but they swear they saw some of them as well. There were lots of bulldozer tracks around that looked like tank tracks. These bright militia members decided the obvious. They had defied the federal government by sneaking around a government road block and stumbled onto a major top secret United Nations operation designed to take over the United States and eventually the world! They couldn't wait to spread the word to other militia movements full of equally bright patriots. I doubt that Congress knows where the big news about all of the "black helicopters and Russian tanks" came from when some of the militia members appeared on national news stations to claim that a big conspiracy is going on. Some of the militia members still swear that the helicopters, troops and tanks remain secreted away in camp somewhere in Northwestern Montana until the time is ripe for the "Big Move"!

I was getting pretty worn down. Chuck had been there even longer than I had been, although he was on day shift. A major shift in the overhead team was coming up as they rotated personnel around to give them a break. It looked like we would have a choice of taking a day off and going to another fire complex or going home.

Our day off came the Sunday of Labor Day Weekend. I had suggested that I would just as soon go home instead of shipping out for another night shift on another fire. Chuck figured he could use all of the money he could make, what with a new daughter and all, and asked for another assignment. Of course he was young and hadn't been on night shift for the duration.

I had spent the previous night on the line. I found Chuck wandering around camp when I got up. He asked what I thought we could do while waiting for orders. We checked a bulletin board. Buses went to Libby every hour. I suggested that we could go into town and get drunk, like some of the other folks did. I've done that before. There is a down side, especially if you have to go back on the line somewhere the "day after." He noted that there was a church service put on by a National Guard Chaplain later, and suggested that might be a better choice. I agreed.

We attended a very impressive church service, the only one I have ever been able to attend while on a fire assignment.

The Army isn't the only place where you have to hurry up and wait. We still had no word on whether we were being reassigned or going home by bedtime. I turned in about eight p.m. and got a good night's sleep. I awoke the next morning to find that Chuck was reassigned to the Yaak Complex, and had shipped out during the night. My orders were for home, providing I could get my stuff together in time to catch a truck headed for Kalispell.

I thought my adventures were over for that trip. I made the flight to Butte, where I was met by a driver from the BLM. His orders were to drive me home to Dillon.

This was Labor Day, last day of a fair/rodeo/big drunk weekend in Dillon. We were a couple of miles short of Dillon when the driver said "What the hell!"

I looked up to see a big old Ford car speeding north in the south bound lane of the Interstate. The driver swerved directly at us, trying to hit us head on. My driver headed for the ditch. I thought the other driver might have had a heart attack or something to explain his going the wrong way on the highway.

I looked into the driver's seat to stare directly into the eyes of one of the most determined men I have ever seen. He was out to kill us, or at least make us do what he wanted. It obviously didn't matter what happened to him. When my driver swerved off the pavement into the ditch, we obviously had bowed to his will. He didn't miss us by more than a foot.

He swerved back onto the Interstate and continued north, accelerating as he went. There was quite a string of cars behind us and he tried to hit every one head on. We didn't have a radio in the government van we were in.

My driver said "What can we do?"

That was a good question. We couldn't call anyone, and we sure weren't about to catch someone in a big Ford with a V-eight engine with the little four-cylinder van we had. Besides, what could we do if we caught him?

I said "He's going to kill someone for sure. Let's get to town and see if we can find a policeman!"

We met the first patrol car as it screamed onto the Interstate at Dillon. Apparently one of the other victims had a cell phone and called 911. The ambulance was right behind the police car with several police cars and a fire truck behind it.

We arrived at fire dispatch to hear that the Ford was going about 110 miles per hour when it slammed into a Toyota pickup. Miraculously, no one was killed. It turned out that the perpetrator was spaced out on cocaine and booze and was too relaxed to be hurt.

The pickup occupant received very serious injuries that will cause him problems for the rest of his life. The junkie had just chased a van filled with a family with small children into the ditch and came roaring back onto the pavement to hit the pickup on the passenger side. Considering the number of cars he assaulted, we were all very lucky.

Fortunately, the judge wasn't willing to accept that the perpetrator wasn't responsible since he was high on drugs. He was sentenced to twenty years in the state prison when others and I testified at his sentencing hearing.

Joe Wagenfehr, fire coordinator in Missoula, called a couple of weeks later, asking if I could take over any of the complexes. I could choose my position. It was late in the season, and the regular overhead teams were really wearing down. I probably would have accepted the invitation except for a big "western" reception for our son and his new bride scheduled for October 1. Most of our family and friends hadn't been able to make the wedding in Wisconsin, and they would all be here. I couldn't accept the assignment and still make that engagement. We hosted the big reception just before it turned winter.

CHAPTER 45

More Smoke

Whether I could sign on to fight fire for the Forest Service or had to go through other units didn't matter in 1995 since it rained in the West during most of the summer. None of us were called.

Things changed in 1996. The spring continued fairly wet and cool, but suddenly it was summer. No rain fell and it got hotter and drier. Fires started with a vengeance. It wasn't long before agencies started running into the predictable shortage of overhead. Major appropriation reductions significantly reduced personnel in federal agencies. Agencies couldn't come close to getting everything done that the public was demanding before the reductions started. Now it became impossible to free a person up for necessary training and fire assignments when they are trying to cover the work previously done by two or three people.

Lois and I had our own personal business to establish.

The guide service business was tough to market, and was slow to develop although things were picking up, but the land management consulting was doing pretty good. We weren't getting rich, but it kept us busy.I had to turn down the first couple of calls because of consulting commitments.

Fall came and I still hadn't been able to make it out on a fire. Then the normal middle of the night call came in early October. A car ran off of I-90 just north of the Wyoming line on the Crow Reservation and rolled several times. It caught fire and high winds drove the fire east through dry grass. Thousands of acres were on fire and several ranches were threatened. They needed a Planning Section Chief. I didn't have any conflicts. Lois joked that it would be all right as far as she was concerned, providing I didn't re-kindle the affair with Rosie Yellowshirt and spend the rest of the fall on reservation fires. That wasn't about to happen since getting home in three weeks for opening day of elk hunting season was more important.

Morning found me flying to the Rollover Fire south of Crow Agency with Hal Wetzsteon, another Beaverhead retiree, who was going as Incident Commander. It felt good.

We flew into Billings, Montana, met other team members, picked up some vehicles, and headed for Crow Agency. The fire had burned close to 10,000 acres by the time we arrived. Tom Corbin, Forest Manager for the Bureau of Indian Affairs, led the briefing. There were lots of old fire friends on hand. Randy Pretty On Top was acting as Incident Commander until we could take over. We had been there before.

The Rollover Fire was a rather typical late season Eastern Montana fire. The primary fuel was grass with heavy brush in the draws and wet areas. The fire burned into some ponderosa pine, choke cherry, green ash and other "brushy" fuels higher into the Wolf Mountains east of I-90. The initial attack forces had done a fantastic job, considering the high winds and dry fuels. They obviously had some exciting moments. Miraculously, no major structures were lost although a few sheds, miles of fence, some hay stacks, a lot of grass needed to sustain livestock and other resources burned.

The wind had died down by the time we arrived but extreme winds of over 50 miles per hour were predicted. The fire would take off again if we didn't get the fire line completed before the wind arrived.

We had good crews and equipment. It was a matter of getting what we needed in place and getting the line completed and burned out before the winds picked up. That looks pretty simple on paper but it's another matter on the ground.

The late season presented additional concerns with the potential for significant weather changes, and a significant weather change was predicted. We established the Incident Command Center in a conference room at the agency dispatch center. Computer assistance, word processing facilities, copiers, and good communications were available. We even had a FAX to relay information.

We housed crews in buildings at the Powwow Grounds, and engines and other water handling equipment in various garages around the town where they wouldn't freeze. Engine crews were housed in the Catholic Church.

It's hard to track and coordinate everything when men and equipment get scattered out like that. It can work when experienced people are involved. We had experienced people. I had good help in Planning's Resource Unit from a rancher's wife who maintained her fire qualifications after leaving government service to give her family and ranch work more attention. She kept good track of the resources we had.

The weather gave us a few breaks and we caught the fire in a couple of days. The wind arrived, but so did some rain with snow on the problem section. Now the primary concern was how to get the crews and equipment where we needed them. Eastern Montana includes some soils that become extremely slick when wet. We had slick, wet soil and some pretty cool weather. Fortunately, the wind blew pretty hard and dried things enough so we could get around.

We got the fire well into mop up and turned it back to the agency in plenty of time for me to make my planned elk hunt.

CHAPTER 46

The Sunshine State

We spent another wet year in Montana in 1997. I took the needed physical tests, just in case, but there was little action and no calls.

The media started to focus on fires in Florida, Texas and other southern states in 1998.The calls started to come, most of them involving Planning Section Chief assignments in Florida. Conflicts with our consulting business persisted so I turned the first couple down. Then a call came that I could accept starting June 25. There wasn't anything that couldn't be rearranged until July 10, when Lois and I were scheduled to look over some major watershed problems for a client in Canada. Besides, I had never fought fire in Florida. They agreed to have me back in time for the July 10 trip to Canada. I was on the next commercial flight out of Butte, Montana, heading for Jacksonville, Florida as an unassigned Planning Section Chief.

I met Jim Freestone, Fire Management Officer on

the Wise River Ranger District, Beaverhead-Deerlodge National Forest, at the Butte airport. Jim was headed for Jacksonville, Florida as an unassigned Safety Chief on the commercial flight I was on. Jim presented an interesting theory. Rod Bullis, Minerals Forester on the Lincoln Ranger District, Helena National Forest, had been dispatched as a Type Two Incident Commander a few days earlier. It sounded like fire officials in Florida were trying to assemble a Type Two Team on site. Rod had been an Operations Section Chief on one of the east side teams for me while I was Incident Commander in the mid-1980's, It seemed like as good as any rumor I could come up with so we passed it on to other fire fighters who joined us at various airports along the way. Our plane picked up Roger Gowan, a Forester on the Gallatin National Forest in Bozeman. Roger is a qualified Finance Section chief. The rumor we developed seemed even more feasible so we kept passing it along.

At various points, including Jacksonville, other personnel necessary to assemble a Type Two Team appeared. Jim Steele, a Bureau of Indian Affairs Forester in northern Idaho, joined us as an unassigned Operations Section Chief. So did Gary Cole, a Forester with the U.S. Forest Service in northern Idaho, as a Logistics Section Chief. They were all good hands and I had served with them before. We were now short only an Incident Commander and another Operations Section Chief (a position never filled this trip), and we had a team.

We arrived in Jacksonville, Florida in the middle of a very hot, humid (to a westerner), smoke-filled night. A State of Florida Forester picked us up at the airport and hauled us to a, thankfully, air conditioned motel. All he knew was that we were heading for Baldwin, Florida the next morning and had some vehicles to drive. A slight language barrier became immediately apparent. When someone in Florida mentioned "far", they usually were talking about what we came to Florida to fight, not the distance we had traveled. I also determined that "jeetyet?" was simply an inquiry as to whether we'd been fed recently. "Sup" was an inquiry involving what we were doing at the moment rather than an abbreviation for supper.

The vehicles arrived shortly after 6 a.m. that morning. Someone passed us a map, and we headed west. Thank God I didn't have to drive in that urban traffic. Those who volunteered to drive had obviously spent far too much time driving mountain roads in Montana and Idaho. I don't know why we didn't cause a dozen accidents; it definitely wasn't our fault. Maybe Florida drivers are just used to tourists doing dumb things. I was still praying when we arrived at Baldwin.

A few fire personnel remained at Baldwin Police Station. They reported that we had just missed the Big Move. The fires in that vicinity were under control so we were actually headed for the Withlacoochee Education Center somewhere miles to the south. They handed us a map, we got back in our vehicles, and continued to terrorize traffic through a fair part of Florida. Firefighting is similar to any campaign situation, like armed services. Mostly you hang on and hope you end up in the right place at the right time. We hurry a lot, wait a lot, and work a lot when the time comes. We pulled into Withlacoochee (and several places we thought were Withlacoochee) shortly after dark, having left a long trail of near accidents along the way.

We found that the rumor we had been spreading was

true. Bullis met us at Withlacoochee where we were briefed by a State of Florida Type Two Team that was in place. They were a great team that was used to fighting Florida fires. However, they had been on the line for 21 days and needed a break. We were to work with them for a day, then take over their fire complex (something like 18 fires scattered over a large part of central Florida). They would take their necessary time off and then return to take the assignment back. It was a smooth transition, and everything worked amazingly well.

Planning isn't my favorite activity, but I had some good help on Withlacoochee. We'd debrief personnel coming off the line, analyze fire line radio traffic and establish priorities. We worked with the other units and developed necessary shift plans.

Distance between fires and the fact that most forces on the line simply went home when off shift, rather than returning to a centralized camp where we could interact, complicated things considerably. In many ways we simply prioritized who got what, and on-the-ground action on individual fires usually progressed fairly independent of what we planned.

We coordinated daily with the Interagency Command Center in Tallahassee to advise them of our situation, needs, or surpluses. This information was used to establish priorities for incoming resources. Reports had over 2,000 fires burning in Florida, eventually covering over 500,000 acres in 67 counties.

We had our hands more than full with the multitude of fires in our complex with our skeletal overhead team and predictable problems involved in managing personnel with whom we had minimal contact. I personally wasn't familiar with all of their fuels, and have had very little experience fighting fires in areas that were that flat and populated. We furnished the basic organization and supervision needed to run the fire and depended a lot upon local firefighters for details.

Details required new adjustments in our thinking. When someone mentions a ridge, I start looking for a noticeable difference in elevation. In Florida, a ridge is wherever the soil is sufficiently sandy and dry to grow pines. And yes, it might be a foot higher in elevation than the surrounding hardwood jungle. A "bay" is a wetter area, generally supporting bay trees. Underbrush was everywhere, ready to burn, and it burned well. Most of Florida has burned historically on a reasonably short time-frame, like every 10-15 years. Fire frequency kept fuel accumulation low and helped create ideal conditions for wildlife and watershed. With the dense human population, subdivisions, intensive agriculture, commercial tree farming, etc. fire wasn't allowed to play its natural role. The brush and other fuel accumulation had reached a critical mass. Now they were paying.

Pine plantations in Florida reach saw timber size under intensive management in 20-25 years compared to ±100 years in the Northern Rockies, which intrigued a Forester like me. The pine plantations burned very well. Involved industry would be hard pressed to salvage and process all of the trees we burned.

We worked hard and made progress on Withlacoochee, finally getting everything controlled except for one fire in Lake County where the ground was too wet for line construction with standard equipment. We had a "muck fire." Local forces reported that the perimeter was lined, but chances were that it would linger on and eventually escape to cause new problems. Muck fire was a new phrase to me. A local firefighter explained that some big "Yankee" developer had drained a swamp back in the 1950's, with the intent of "planting Yankees." I decided that meant a rich realtor probably received federal aid to drain a swamp in the 1950's, with the intent of constructing retirement homes or condominiums for rich people from the north who were tired of cold winters. Apparently this process became too common and a lot of the developers, including this one, went bankrupt. He left a drained swamp behind. The resulting top layer of soil is all organic material in various stages of decomposition for several feet deep.

Now I understood. We face the same problems with peat fires in the north where accumulations of organic material can be several feet deep. Fire smolders in this material and will work its way deep into the ground and eventually pop up where least expected. These fires are almost impossible to put out unless saturated to the depth of the smoldering material.

I suggested that we install a sprinkler irrigation system through the problem area. We could blanket the "muck" with sprinklers and soak everything to the depth the fire had reached. I'd used irrigation to put out peat fires in the past, so it should work on muck fires.

Local forces liked the idea but lacked an adequate pressurized water source. We discussed drilling a well with a local well driller. He informed us that he couldn't do it for at least ten days, would only do it for a ridiculously high fee; and besides, we needed a state permit and that took months. The permit wasn't a problem. Florida's Governor Chiles had decreed that such things as permits weren't necessary for temporary activities designed to help control the mass of fires that plagued his state. The ten day time frame and exorbitant price demanded by the local well driller caused heartburn for someone used to seeing reasonable action. The governor had activated the Florida National Guard. In fact, a National Guard captain was with me when we got the well driller's input.

I turned from the well drilling contractor and asked the captain if there was a National Guard unit with well drilling capability in Florida. His response was admirable and immediate. "There is, sir. I can have them activated and on site in less than 24 hours, sir. We can have two six inch wells deep enough to supply your water needs and powered by a pressurized system ready to go in eight more hours, sir. We'd be all too happy to provide you with that service, sir."

Now we were cooking although I wasn't real used to being called sir, having only made Specialist Fourth Class during my two years in the Army.

Anyway, he had the response I needed. Then the politicians got involved. We returned to the command center where we received a phone call forwarded from the governor's office. The private well driller had called the governor.

He claimed we were trying to put a poor honest well driller out of business with improper use of taxpayer money. He had volunteered to do everything we asked immediately and was asking only enough to cover his expenses. We insulted him by saying we would use the Guard instead of a reasonable private business that had volunteered to set aside his regular busy schedule to meet our needs. We were politically forced to call off the Guard to quiet the politics down, and got our wells. I'd have loved to tell that lying well driller to shove his well where the sun doesn't shine, but the choice wasn't mine to make.

Between planning sessions I managed to get in some helicopter time to see what we were up against. It's tough to plan a campaign if a person simply depends on what others are saying. A lot of the fires in the complex weren't all that large, most caused by lightning. Florida obviously has some horrendous lightning storms. It was tough to tell where one fire stopped and another started in some areas. And houses were everywhere. And then there were the problem fires. I lost my brief case on the return trip, so I have no plans to check fire names and key personnel. There were lots of them.

Firefighters depend a lot more on mechanical equipment to combat fires in Florida than we do in much of the West. The terrain and access is much more favorable. It rapidly became obvious to me that the heat and humidity makes mechanical control much more critical. Heat stroke isn't all that uncommon with hand crews in the West, but it's a constant companion there. Most fire line is constructed with special plows, frequently pulled by bulldozers with extra wide tracks for working in swamps. Water is common, and with the level ground, dense population access for fire trucks is critical. Control of the airways was a new problem for us as well. We are used to closing the air space over fires to make it safe to use retardant planes and helicopters without having to worry about running into an airplane we don't know is there. Here we had both heavy civilian and military aircraft overhead almost constantly, making closures much more difficult.

I managed to get on the ground on the muck fire, and

a fire where the swamps were delaying line construction just to the north. The irrigation system wasn't in place yet, so we were spending a lot of time trying to supply water to support small four-wheel drive fire engines that were keeping the fires from spreading. We spent a lot of time getting the trucks unstuck.

The original Florida overhead team was back in place by this time. We received orders to pack up our gear and head back to the Baldwin Police Station near Jacksonville. New fires had taken off in that vicinity. We were directed to take over the new complex. We dug out the old map, jumped in our vehicles and terrorized traffic north.

Smoke and fire dominated the whole countryside as we approached Baldwin. Local crews were doing everything they could but were losing ground fast. Bullis and Steele headed out to size up the situation on the ground while the rest of us tried to get the command post established.

A new team member joined us at Baldwin. Allan Gierrer, a member of the Salish Indian Tribe out of Polson, Montana, was assigned to us as a facilities manager. I love working with people like Allan who was all team and all go. Fire personnel must be capable of making decisions fast when problems develop. If someone else hesitated on a decision, Allan made it for him. If his answer headed us in the wrong direction, he was perfectly willing to try a new solution. Indecision under these circumstances can kill, so he was a welcome addition.

We had unmanned fires to the north that had cat lines around then and weren't moving that much. Our major concern was that they might escape control lines since no one was there to keep them contained. While Bullis and Steele were out flying the fires to the south, we located a couple of National Guard crews standing by just south of us. They had necessary training and were more than anxious to do what they could to help.

We were concerned about putting them up against running fires before they received some on-the-line experience. We really didn't have the overhead to help them work a hot line effectively. The fires to the north weren't moving that fast. The Guard troops were more than ready to take these fires when we contacted them. The fires to the south were another thing altogether. We could see the huge convection columns from Baldwin. The radio traffic sounded even more ominous.

We had to set aside our usual planning and implementation phase to simply help get a massive evacuation program underway while trying to save as many houses as safely possible. Things did not look good. Bullis and Steele confirmed that the situation was beyond our capability as an understaffed Type Two Team as soon as they returned from flying. We learned that a fully complimented Type One Team from Utah, southern Idaho and Nevada under an Incident Commander named Monahan was available. We advised the unified command center in Tallahassee that we needed Monahan's team as soon as they could get there. We still had to manage the fire situation for the better part of two days. Thankfully the Guard troops kept things under control to the north. In the south, most of our time was spent just trying to keep people out of harm's way while trying to save houses. A whole county was evacuated. In sheer numbers, I understand that was the largest evacuation of people in U.S. history in the face of running wild fire at the time.

The media couldn't figure out why some houses burned

and others didn't although it is simple: some people do an outstanding job of managing their land. They keep the brush under control, frequently by using prescribed burning or mowing, and their trees are well pruned. Litter was not a problem. Only low intensity fires could move through these areas, if they burned at all. These houses were saved. Others apparently like the isolation provided by dense underbrush. They have brush, trees and debris right up to their houses, rain gutters full of dry leaves, etc. Their houses burned.

One evening in the middle of all of this confusion I found time to pick up a cell phone and call home from a pickup truck in the parking lot. Lois was happy to hear from me, but stated that a favorite cousin, Robert Pence, had called from Spokane, Washington to say goodbye. He had lung cancer, the somewhat predictable outcome of many years of heavy smoking, and didn't have long to live. She gave me his number and I called him.

"So what's going on," was his somewhat nonchalant inquiry when he found out it was me, as if everything was routine.

"Oh, I'm just sitting here in a pickup truck in Florida crying because I've heard some sad news," was about all I could choke out at first.

We went on to have a good discussion. He seemed to be taking it all a lot better than I was. He was a special guy. I hope I can handle it as well when it's my turn. I figure that a wildland firefighter has spent at least half his life walking uphill. I hope I'm still walking uphill when the time comes.

We got Monahan's team in place. They were more than welcome relief and I knew a lot of them. Some were old fire dogs like Wayne Smetanka, who came out of retirement in Montana like I did to join the team.

Our next assignment came with Monahan. President Clinton had authorized use of the active armed services to assist the fire effort. Heavy firefighting equipment from all states in the nation was being activated and sent to Florida. Much of it was being loaded on huge Air Force C-5A's in California, Oregon and Washington. Someone had to set up a staging area to off-load this major air lift and see that everything got to the right destination. We were elected.

We were handed a map and took another terrifying dash through urban traffic. Fate was kind, and we survived to find Jackson Naval Air Station (JAX) in Jacksonville. We arrived late on a Sunday afternoon. The officer in charge met us at the gate. Her normal assignment was dining facilities (mess) officer, which was a big help later. The base personnel were expecting us and we were escorted to their structural fire station. We were briefed by Navy officers and base fire personnel concerning what they knew at 6 p.m. The first flights were in the air, with expected time of arrival about 10 p.m. Media had been alerted for a major media event. Time was of essence.

We discussed who needed to do what. We needed incoming flights to report to the aerial fire station, which necessitated our crossing a "hot" landing strip with crews as they arrived. Due to media involvement and political opportunity, the first flights were scheduled to arrive in front of the flight control tower on the base side of the strip.

The media would have a chance to photograph the first three C-5A's as they came in from California, record the base commander and politicians welcoming them to

Florida and interview crew members. After this event, our job was to get the equipment across the "hot" strip to the aerial firefighting station where they would be staged until we could get involved firefighters fed, sufficiently rested to meet safety requirements, and dispatched to their assigned duty stations.

The flight control tower tracked the incoming flights and had a rough list of what crews and equipment was on each the plane. The list even included the fire incident command center where they were assigned as soon as we could get them there.

All fire engines and other motorized equipment had all but enough fuel to get them on and off of the planes drained before they were loaded on the C-5A's for safety reasons. Their drivers would drive them off the planes and park them in a specific area behind the station.

Bullis would welcome crews, give them a quick outline of what was going on, and direct them to a specific door in the station. He also had to coordinate everything and pick up the pieces. I would be waiting at a desk inside the door where I could confirm which crews and equipment had arrived, where they were assigned, and if they needed rest before being sent on. Freestone would assign barrack rooms for those needing rest and see that they received a box lunch. Gierrer would escort the now fed fire fighters to the barracks by bus, where Cole was getting them assigned and rested. He was arranging a wake up time. Crews would get fed at the mess hall then be bused back to their equipment. While the crews were resting, the base firemen had arranged to get their equipment fueled. Base shore patrol would escort the crews to a side gate where they would be met by law enforcement personnel who would "convoy" the fire equipment and crews to their assigned duty station.

Once they cleared the Jacksonville area traffic, things should go pretty smooth, since the major interstate and other highways south of Jacksonville had been overran by fire and closed to civilian traffic. Gowan would handle the financial end of things-trying to keep us all out of jail for misuse of firefighting funds plus handling contracts, claims and other fiscal challenges.By the time we had all of this figured out the planes were only a few minutes from landing. We made a quick dash to the appointed place where all sorts of media, Navy personnel and political figures waited in a mass of confusion. Three huge air craft dropped out of the dark, smoky sky with a roar, taxied to the right spot, and dropped their loading ramps. The very important people went on board and welcomed the first of many. The aircraft carried between five and seven fire engines (depending on size of the engines) and their crews on board. The media had a field day photographing the equipment as it drove off the planes with the little fuel left in them. They interviewed as many of the incoming personnel as they felt was necessary. Appropriate speeches were made, although probably not heard over the whine of huge aircraft engines. The crews climbed on their equipment and we led them across the landing strip to where we needed them. Then the planes were back in the air, heading west to pick up more personnel and equipment.

After a few minor bugs, the check in process went very well. It didn't take long to determine that most of the incoming personnel had been on duty for over 30 hours. They all insisted they had enough cat naps on the planes and were ready to go. They came a long ways to fight fire, and that's what they wanted to do. Safety requirements dictated a minimum eight hours of rest before they could proceed.

Another problem involved separate rooms for male and female crew members. The Navy wasn't interested in having the media make a big deal out of having crew members of different sexes sharing the same rooms, although they would be sleeping in clusters of small back-pack tents on the fireline. Neither were we, and most of the crew members for that matter. That put a major hitch in trying to sort out who was assigned which room. We had a lot of people backed up who needed rest, so we didn't have much time.

Freestone made a valiant effort to track who bunked where for as long as he could. He eventually just assigned rooms to crews, told female crew members to get together with other crews and crew bosses on who was actually in which rooms, and make arrangements to be woke up with the rest of their crew. There wasn't time to do otherwise.

And the planes kept coming. We'd just get one batch of personnel and equipment processed and another flight of C-5A's would come roaring in. It all started to run into a blur. Things became mechanical after a while. It went that way for 36 hours straight. After the first eight hours we weren't only greeting incoming flights, but also had to deal with rested crews heading out.Everything worked surprisingly well.

By this time commercial airlines were flying numerous twenty person hand crews in from various points in the west. Most of these came in through Jacksonville International Airport where we had to pick them up, get them processed and on the way to assigned fires.

There is no way that we could have handled that load

with just the personal we had on our team. Jacksonville Naval Air Station became a critical part of our team. The Navy and related civilian personnel was there to help, and help they did. Base personnel already had full time jobs. They still managed to give us any assistance we needed and we needed a lot.

We just got to where we could see daylight when a major flight of totally unexpected (by us) National Guard C-130's and other aircraft landed.

The new arrivals involved equipment and personnel from Virginia under EMAC. EMAC stands for something like Emergency Management Assistance Compact. The governors in the Southeast had reached agreement that they would exchange assistance as needed for emergencies. The agreement included National Guard coordination. This was the lead flight of many to come.

The new flights had been coordinated through the Federal Emergency Management Agency (FEMA). Someone, somewhere apparently assumed that since the other flights were going into JAX that was where these were to go. I'm sure someone, somewhere knew what was going on but we had somehow gotten left out of the loop. FEMA eventually caught us up on the situation and asked if we could handle it. We did. Actually, a lot of this new equipment arrived by rail. We had to make some major shifts to work in the civilian railroad depot, which was some distance from the base, but it worked. We processed a lot of state, local and contract EMAC equipment and personnel from Virginia, Kentucky, Tennessee, the Carolina's, and probably some other states.

All personnel and equipment were supposedly checked for condition and qualifications before leaving home base, with questionable results. Some incoming personnel appeared more than a bit out of physical condition to meet requirements, and others looked very young. I think some of those good old boy rednecks brought their under-aged grandkids along for the employment with little serious thought that actual work and danger might be involved.

We eventually started to catch up. Bullis called to ask if I could come to the new incident command post. We had been so busy that I missed the move from the structural fire station to the Navy's Emergency Command Center.

One of the immediate emergencies was that we had been so tied up processing equipment and personnel that I had not submitted a required Incident Command Form 109 to the Interagency Coordination Center in Tallahassee for over two days. The form is essentially a status report on what is going on at your incident. It includes a summary of your situation and the personnel and equipment currently involved.

That's a no-no for a Planning Section Chief, and the concern was valid; paper work doesn't go away. Although we had been over our heads just getting everything where it had to go, Tallahassee really did need to know what was going on. Besides our "incident" was simply a staging role, so essentially everything we had was destined for somewhere else. We had verbally been trying to keep them advised, but now we had a chance to catch them up. I hammered out an IC-109 for them.

The new command center was impressive: air conditioned with good phones, computers, a FAX and other accessories that usually aren't so handy on a fire assignment. I think it was designed to survive a direct hit from an atomic bomb.

We were also processing a lot of unassigned overhead personnel from the west by this time, most arriving by commercial flights without a specific destination. We checked their qualifications and advised Tallahassee.

At 1700 (5 p.m.) each afternoon, all Planning and Logistic Section Chiefs on every incident in Florida were involved in a conference call managed through the unified command center in Tallahassee. Personnel at Tallahassee gave an overall situation report which primarily covered how many fires were burning, where the priorities were, and what the weather was doing. Also included were any political or media problems. Then each incident gave a status report, including what they needed and what they could release that might be of value elsewhere. We were usually called on last since we had the best opportunity to match incoming personnel and equipment with needs on a fire somewhere. It worked well and we seldom had anyone or anything standing around for long.

We had to be innovative at times. We received several relief tractor/plow operators out of Tennessee, sent to keep the tractor/plow units working 24 hours a day. We had no calls for them. However, there was a need for Tractor Bosses. Tractor Bosses work ahead of tractors flagging where fire line is needed and to monitor time and keep them out of dangerous areas. These people qualified so we sent them out as Tractor Bosses.

Several Strike Team Leaders arrived from Nevada and Idaho. Most of them worked for their state forestry departments. I knew lots of them. The rest either knew one of my brothers or our son or daughter.

One of the Nevada Division of Forestry Strike Team Leaders made the news a couple of nights later. A fire was making a run towards several homes. A TV reporter interviewed a frantic homeowner. Apparently he had tried to get a volunteer group from Florida to help save his home. The volunteers were not trained and weren't making progress. Suddenly the Nevada Strike Team Leader showed up and took charge. The Strike Team Leader got his crews in place and the structures were saved. The home owner report was "Those folks from Nevada sure know how to fight fire! I wasn't getting anywhere with those guys from Florida." The fact that didn't come out was simply that the volunteers lacked training and leadership. The organized group had both.

A young lady from California eventually showed up to take over my Resource Unit Leader position. She knew what she was doing, which really took a lot of pressure off of me.

We picked up motel rooms (with air conditioning!) and settled in to a more manageable work schedule. We got up about 5 a.m., ate breakfast at the Navy mess hall, went on shift at 6, ate lunch around noon (depending on what was going on), went off shift somewhere between 6 and 8 p.m. (again depending on what was happening), retired to the motel, ate supper, had a couple of beers and went to bed. Life was becoming marginally routine.

The assignment continued to be challenging. The tribal council called from the Rocky Boy Reservation in Eastern Montana. One of their crew members managed to slip away before taking the required pack test and annual required "Standards for Survival" training, although he had plenty of experience. He knew the training was mandatory and they planned to make an example of him. Did we know where he was? We did. He was to be picked up and returned to Montana on the next commercial flight. The cost of his round trip plane ticket would be subtracted from his pay. We just got that one settled when we got a call from the Blackfeet Tribal Council at Browning, Montana. One of their experienced Squad Bosses was also guilty of missing the two requirements. We rounded him up and sent him home. Then we got word through both agencies that they understood a U.S. Forest Service ranger had gotten a fire assignment in Florida without having completed either requirement. What were we going to do about her?

We were able to track her down as well. Apparently her husband was a smokejumper from Idaho who was assigned to the unified command center in Tallahassee. She was a ranger in Montana. They hadn't seen each other in quite a while. She was qualified to fill a needed office position in Tallahassee, so she volunteered although she had not completed the pack test or annual training. The Bureau of Indian Affairs people wanted to know why they had to pull their people while the Forest Service didn't seem to care. In reality, her assignment was far removed from the fire line, so one could logically question the need for the physical test or safety training. The Montana Indian Fire fighter crew members were in line assignments. She stayed while the crew members went home. The decision was not mine to make.

Several line personnel shared our motel so we heard some of their stories each evening. The Tractor Bosses from Tennessee were always good for a few stories. They were classic "rednecks" with some interesting views. Our primary problem with them involved persistent "passes" at waitresses that left the whole operation open to sexual harassment claims.

By this time, our team was starting to run up against the twenty-one day limitation on how long we could be on duty without a couple of days off. Disney World and other Florida attractions were offering free passes to firefighters if they wanted to take a couple of days of Rest and Recreation in Florida, then go back on duty. It was an attractive offer for most of the team.

I was facing my self-imposed deadline of July 10 for making the consulting trip to Canada. I really couldn't miss it. I started things rolling to get me home as promised, and it actually worked. I flew out of Florida on July 9 and made my Canadian commitment.

CHAPTER 47

The Fires of 2000

There apparently wasn't much need for retired firefighters in 1999, although it was a dry year. The drought continued into 2000, and a whole new fire season came with it.

Wildland fuel accumulations appear to be reaching a critical mass in many areas. Areas with significant fire history need only the right burning conditions for major catastrophic fires to occur. The burning conditions arrived early in the northwest in 2000. Major fires were burning by mid-July. Smoke filled the sky for days, and fire personnel were being dispatched as fast as they became available.

One evening, the news reported that lightning had started a new fire on the Clear Creek drainage on the Salmon National Forest in Idaho, just west of us. Most fire teams were committed elsewhere. The Clear Creek Fire was in the River of No Return Wilderness. Media reports indicted the Forest Service would "monitor" the fire rather than divert forces from fires burning closer to developed areas.

Clear Creek is on the old Cobalt Ranger District where I fought fires between 1961 and 1969, and many had been in the Clear Creek vicinity. Some of the toughest topography in the United States is involved, and God meant for it to burn frequently. Lois and I discussed the fire's potential near the eastern boundary of the wilderness area. It was going to be interesting to watch this one grow.

We didn't have to watch for long. Extreme burning conditions including high winds, low humidity, and high temperatures swept across the region two days later. Ashes were falling in Dillon, approximately 100 air miles east of the fire, by nightfall. Smoke reduced visibility to a few hundred feet in Dillon. The fire swept north, east and south out of the wilderness area. Thousands of acres burned in one afternoon, including a barn on Panther Creek. Other structures, including an expensive facility that had been built at tax payer expense to treat toxic waste from mining activity at Cobalt, were threatened. Clear Creek moved from monitoring status to a national priority fire in just a few hours.

Fires were burning in our immediate area as well. Crews were essentially all assigned somewhere. They would just return from one assignment to be sent on another. Our oldest daughter, Jan, now works as litigation and appeals coordinator and paralegal for the Beaverhead-Deerlodge National Forest. Her husband, Chuck, resigned his job as Forest Service wildlife biologist to start a ranch management consulting partnership. Chuck tended their three small children while Jan fought fire.

Dick Owenby, Fire Staff Officer, called to see if I could

help. Personnel and equipment were in short supply. They needed someone who could survive observation flights through rough air in a single engine air plane without getting air sick, pick up fires in the very smoky terrain, radio in the location, determine priorities and what was needed for initial attack on the fire, and guide crews to fires in a safe manner. We had several prior commitments, including the pending arrival of a fifth grandchild and several clients interested in taking scenic tours along the Lewis and Clark Trail. Dick assured me that we could work around those commitments. I signed back on in a contract position and went flying.

Dennis Devivo held the contract to fly the daily patrols with his Cessna 180. We made a good team. Flying through mountainous terrain a few hundred feet above the ground following thunderstorms isn't for someone with a weak stomach. I've ridden bucking horses that offer a smoother ride.

We located and reported a lot of fires. We'd see the smoke column from a fire, Dennis would fly over it and we'd radio location to the fire dispatcher using coordinates shown on the plane's Global Positioning System (GPS). I also knew the local terrain well enough to use landmarks to report the location by drainage name or mountain, and advise crews concerning the best route to take to reach them.

Then Dennis would take the plane into a tight spiral around the fire at about 500 feet above the ground so I could size things up. I'd give the dispatcher my best interpretation concerning fuel type, current fire activity, and estimated fire potential. I'd give them my best guess concerning what was needed to control the blaze if it warranted immediate attention. This last report was all too often a pretty futile effort. The personnel and equipment simply was not available to take on everything we found. A national priority was to hit all new fires as hard as was necessary while they were small. There were all too many huge fires going on in the west. We didn't need more, but we could only stretch things so far. Fires to the west continued to spread. Smoke choked the sky, especially during the afternoon. Flying conditions frequently were so hazardous that we simply could not see.

Dispatch called one evening. We had completed our routine patrol and had basically gotten smoked out of the sky as fires in Idaho and beyond went crazy that afternoon. A severe lightning storm with essentially no rain was pounding the upper Bitterroot Valley and was moving across the Pintler Range along the Continental Divide that forms the north and west rim of the Big Hole Valley. A fire had been reported in the Mussigbrod drainage north of Wisdom. Could we check things out?

It was a good question. Dennis was waiting when I got to the airport. We launched into the smoke filled sky and headed towards Wisdom. Visibility decreased rapidly until the ground essentially disappeared as we approached the divide separating the Beaverhead and Big Hole drainages. I have a specific allergy to flying around mountainous terrain in small single engine air planes at low elevation when I can't see the ground!

Dennis probed west, trying to find a hole in the smoke. It wasn't there. A heavily timbered ridge suddenly emerged from the smoke, rising immediately in front of us. I recognize the ridge from years of elk hunting. We were not where we wanted to be under those conditions. Dennis banked into a steep turn back the way we had come. The probability of ending the flight wrapped around a tree appeared much more likely than getting through to the Big Hole Valley. The ability to see a fire if we did get through seemed to be decreasing by the minute.

We discussed the situation with dispatch. A jumper plane was available with six jumpers in Missoula. They would try to reach the fire from the north. They had radar and a lot of instrumentation that we lacked. We returned to Dillon with plans to fly shortly after day light the next morning.

Darrel Schulte, Forest Fire Management Officer, flew with us the next morning. Six jumpers had hit the fire just before dark. Dense smoke had settled into all of the canyons during the night. Things seem a bit weird on flights when all a person can see are the higher mountains and ridges poking up through a blanket of smoke.

Dennis Havig, Wisdom District Ranger, was waiting at the grass strip that serves Wisdom as an air field. Finding it was going to be a problem. Our plane was equipped with a GPS that allowed us to fly directly over Wisdom. It had to be somewhere down there through all of that smoke. We were free of the immediate mountains, so Devivo put the plane in a spiral and down we went. The first thing I noted was Mud Lake, a small pond northeast of Wisdom and closer to the mountains than we really wanted to be. At least we could follow the highway to the air field. The most interesting feature for the Wisdom strip, as far as I'm concerned, is the numerous ground squirrel and occasional badger holes. We came to a bumpy stop in front of Havig's truck.

Then it was back into the air and a fast trip towards the

Continental Divide. The smoke plume for the Mussigbrod fire was quite visible long before we got there. The fire was tucked into a dense jungle of timber just on the Bitterroot side of the divide: a bad place for a fire. There were also numerous other smoke plumes in the vicinity that had not been reported. Mussigbrod was obviously not the only fire that the lightning storm had started.

Chatter essentially clogged the radio frequencies common to aircraft. It looked like the entire upper Bitterroot Valley was on fire as we looked to the west. We could see retardant bombers making drops on fires that were already making significant runs up slopes near Sula in spite of the early hour.

We established radio contact with the jumpers on the ground at Mussigbrod. They jumped just before dark and had not really had a chance to size things up on the ground. I assumed that that meant they hadn't had a chance to size it up in the daylight, but had spent most of the night building line and taking other control actions. At least that is the action I would have taken on such a fire during an initial attack effort if I arrived at dark. It wasn't the correct assumption. Current firefighting orders based on safety concerns restrict the number of hours that a firefighter can put in even during initial attack, with specific orders against working on such a fire at night. They jumped, collected their gear, made camp, slept, and were starting to size up the situation facing them as we arrived.

We advised them of a significant spot fire that was flaring up on the Big Hole side of the divide. Two small lakes were located close to the fire. We made arrangements for a helicopter equipped with a bucket to help the jumpers. They'd found a good water source for a pump in the creek below the fire. Arrangements were made to long line a pump to the jumpers.

Our attention was now diverted to other smoke columns to the north along the divide. As I recall, we reported nine fires that morning. One was located next to a logging road and an old logging clear cut. An initial attack engine (fire truck) could reach it. One was dispatched. Two others appeared somewhat isolated by rock outcrops and had to be assigned low priority. The others all appeared to have significant potential. One was already starting to crown and move up slope. Six more jumpers were available and we requested that they be dropped there although it looked like it was more than they could handle. My hope was that they could hold things until additional help could arrive. No one was available to take on the other fires. They would just have to wait. Most of them didn't.

I would like to have continued north along the west side of the Continental Divide on the Bitterroot National Forest and onto the Phillipsburg District on the Beaverhead-Deerlodge since that's where even more fires were popping up. The Cougar Creek fire on the Phillipsburg District was really boiling and major air activity was involved. We could see some of the planes and helicopters involved, but there were others that we couldn't see because of the smoke. The radio frequencies involved were so busy that we couldn't intrude without posing a danger that we couldn't risk.

We swung back onto the Big Hole side of the divide. A group of recreational horseback riders had been spotted in advance of the Cougar Creek fire. Radio traffic indicated that a jump plane was being diverted to see if jumpers could help guide the riders out of danger.

We flew back over Mussigbrod. The jumpers there

reported that things looked tough but they thought they could hold with the water drops and pump. The helicopter couldn't fly to drop water as long as we were overhead. We had a lot of other terrain to check. We left them and headed east to check the rest of the forest.

The Musigbrod fire blew up that afternoon, burning several thousand acres onto the Wisdom District before dark. The six jumpers had to run for their lives. We couldn't get back to check it from the air since smoke from numerous fires essentially eliminated visibility for planes like ours.

We flew again shortly after daylight the next morning. Mussigbrod had obviously grown into a fire with significant potential. I had walked through some of the area where it was burning a few years earlier. The standing lodgepole timber volume was impressive. Even more impressive was the huge accumulation of dead logs that were piled up on the ground under the standing trees. At least two insect epidemics had swept through involved timber stands since the area had last burned over a hundred years ago. Logs cluttered the ground up to ten feet deep in places. It would be suicide to send crews into fuels like that under the existing burning conditions.

Another fire, one that we had missed the day before, was also showing up in a deep, timber-filled canyon immediately west of Mussigbrod. The two fires would burn together. Nothing short of a major storm would stop them and only wind was predicted. In reality, that whole area was simply long over-due to burn, and it did. We couldn't have stopped it from burning if ten times the number of firefighters had been available.

Things got very busy for me about then. Visibility was

so limited that we could not fly patrols on a predictable schedule. Both fire line and support personnel that could meet specific training standards were all assigned. People were volunteering to help, but they needed training.

Events moved so fast they became a jumble. I still made fire patrols when visibility permitted, but also helped conduct fire schools to get more people qualified. As a result, some of the following events did not occur in the order presented here.

Some support personnel, such as truck drivers, only need a four-hour training course in "Standards for Survival." It is also an annual requirement so fire personnel otherwise qualified needed to take the session to be qualified for 2000. The course focuses on how to size up fire activity. It emphasizes action needed when things look bad, including getting out of the entire area or into a safety zone. If all efforts to escape fail, the course teaches those involved how to deploy and stay in a fire shelter until it is safe to come out. We taught "Standards for Survival" in Dillon, Wisdom and Phillipsburg.

The next step was a full 24 hour fire school for new personnel who could go on the fire line. I teamed up with Frank Russell, a retired fire management officer from Dillon, for our first fire school at the Anaconda Job Corps Center. Eighty young people took the training. We didn't need or want to set a practice fire for them as part of the training. Considering burning conditions, even a practice fire in a relatively safe location was not a good idea. The trained crews would be assigned to a fairly safe location on an existing fire with good overhead direction as their first assignment. They would receive more hazardous assignments as their experience grew.

Firefighters need to be in good physical condition before they can be certified. Almost twice the 80 young people we trained turned out for the fire school but a lot of them flunked the "pack test", a timed endurance walk carrying a 45-pound pack. A lot of the young Job Corps members lacked physical and mental condition needed to pass the test. Most smoked and some had drug abuse records.

One young man who failed complained that we were asking too much. I told him that he could be asked to pack more the 45 pounds for more than three miles on the fire line. I also mentioned that I would have been embarrassed if I couldn't have kept up with a 61 year old man when I was 18. His response was a simple "Yeah, but you care. I don't. Besides I don't have to work." I'm not sure what that says for some of the younger generation. I couldn't recommend someone who "didn't care" for one of the crews. He never made it through the training.

The next challenge involved getting 80 more people from neighboring communities through fire school. These people essentially represented the "pick up" source that was prevalent when I first started fighting fires 44 years before, except that they would have the training and had to pass the pack test.

The media simply advertised that anyone interested in fighting fires should be at a certain location at a certain time with specific equipment (good boots, clothing to last 14 days, etc.). Some larger communities have people shielded by population numbers that see opportunity in signing on. For example, two buses filled up with aspiring firefighters in Butte, Montana. Law enforcement officers then boarded the buses, blocked exits, and arrested four wanted felons who had been in hiding. They saw the firefighting opportunity as a chance to get out of town, make some money, and maybe peddle some drugs, but went to jail instead.

The aspiring firefighters were bused to the "Bender Center," an old Civilian Conservation Corp camp that has been converted into an outdoor education center under Forest Service permit to the University of Montana-Western. It is an excellent location for a fire school. George Johnson, Dillon District Fire Management Officer, had returned from various fire assignments and helped Frank and I with this fire school.

Not everyone who showed up needed the full training. A brother and sister team from the Crow Reservation needed Standards for Survival to renew their credentials. We gave them the training and they headed back to the reservation where they could join a crew. A local school teacher only needed the pack test before he could go out with a contract engine. An Emergency Medical Technician who had been a medic in Viet Nam also needed the pack test before being assigned. I gave them the test while George presented Standards for Survival and they were on their way.

I flew fire patrol between fire schools and related training sessions. The dense blanket of smoke originating from fires to the west impacted fire behavior in the area we had to cover. Fire danger on the Beaverhead-Deerlodge Forest was as extreme as it was anywhere. However, the new small fires we found usually didn't demonstrate extreme fire behavior as long as the sun was blotted out by the smoke blanket. But the smoke blanket wasn't always there. Fires that started before the smoke moved in during the late afternoon, or when severe winds kept the smoke from forming the protective cloud moved as fast as they did elsewhere.

We were in the air one afternoon when we were asked to check a new smoke reported north of Butte. Another detection flight normally checked that area. That plane landed at Phillipsburg and the smoke got so thick before it could take off again that it was grounded.

We had the fire's approximate location and headed north to see what we could pick up. We didn't get that far before another smoke column showed up along the Continental Divide south of Butte. The fire had potential, but there didn't appear to be any significant property values at risk. I noted a green vehicle under the timber just west of the fire as we circled. Forest Service vehicles are normally green. I assumed that someone was already on the fire and tried to establish radio contact. No one answered.

We had only circled the fire a couple of times when we noticed another smoke column near Basin Creek Reservoir, a primary domestic water source for Butte. Now here was a fire with potential, surrounded by houses with heavy fuels and high winds involved. Not only was it surrounded by numerous "ranchette" type subdivision homes and outbuildings but the wind would move it directly towards the southern suburbs of Butte, less than a mile away.

I actually did think about what action to take for a couple of seconds.Anyone who has ever visited Butte, Montana would recognize that such a blaze represented good potential for an urban renewable project in an area that needs such attention. My conscience wouldn't let me do it. We called for aerial retardant, all available ground forces and a helicopter if we could get it. Everything showed up and I probably spent $25,000 in a few minutes.

We knocked that fire down in short order. We found out later that some local residents had spotted the fire that morning. Rather than report it, they rushed to the fire on their 4-wheel ATV's, threw some dirt at it, declared it out and left. They didn't bother to tell the Forest Service or anyone what they had done. Their effort held until the day heated up and the wind increased. It could have been a very serious fire.

We could now turn our attention to the first fire we had picked up. We had a helicopter with a bucket by this time. A beaver dam was in the creek right next to the fire. I asked that the pilot check on the green vehicle before dropping water, since I didn't want to endanger any firefighters who might be on the ground. He reported that it was an old abandoned vehicle all shot full of bullet holes. No fire-fighters were in the vicinity. Interesting things happen near Butte, Montana.

The helicopter could scoop water out of the beaver dam, drop it over the fire and be back over the fire with another load in less than two minutes. He had it pretty well soaked down by the time smoke chasers arrived on the ground.

We flew off to check the original fire north of Butte long before that all came together. This fire was on a rock point just north of a road. We could see two pickup trucks with fire fighters aboard on the road and guided them to the fire.

I was starting to wonder about some of the calls I was making by this time. Some of the fires simply were not going out as fast as they should have with the personnel and equipment we were getting on them. I still hadn't caught up to how significantly new policy was restricting the total work hours and night operations on the fire line.

We were just landing in Dillon one afternoon when we got word that a fire on McCartney Mountain just east of Glen, Montana had been spotted from a nearby ranch. We were experiencing some serious winds so things didn't look good. We were over the fire in a matter of minutes. It had some potential because of burning conditions more than fuels. The fire was only a few acres on a very rocky hillside. Scattered Douglas-fir timber with some sagebrush, mahogany and grass was involved but the fire behavior wasn't all that extreme. It was being sheltered by the dense smoke blanket from fires to the west, which obviously helped. Some ranchers were already on the fire by the time we got there and several smoke chasers were already hiking to it. Forty additional fire fighters had been dispatched from Dillon.

Most of the activity was along the western edge where the fire was backing downhill in grass and sagebrush against the wind. The eastern (downwind) edge that should have been moving had ran into some rocks and wasn't that active. There was a significant spot fire developing in a log beyond the rocks but it looked like plenty of people would be there before it became a problem.

Several rural fire trucks (engines) were trying to help. Terrain prevented driving anywhere close to the fire but we got them into position where they could help if the fire spread. Our next priority was to get the forty firefighters onto the right road since they had driven too far north. We got everyone headed in the right direction then flew back over the fire.

We were in a tight orbit over the fire, guiding people

in and radioing recommendations on suppression actions when Dennis asked "Are we supposed to have a thunderstorm moving our way?"

I was pretty focused on the fire and hadn't been paying much attention to other factors. I looked up to see a very dense cloud approaching rapidly from the west. It was smoke from Clear Creek and Mussigbrod. Both fires had gone crazy that afternoon and the combined smoke was about to shut down visibility. A couple of details needed to be closed off. I asked Dennis to make two more circles over the fire. We couldn't see the ground by the time I got through. Dennis set the plane's GPS for Dillon and we headed home. I'm glad he had the GPS unit. We'd have had a tough time finding the airport without it.

I was anxious to see how things were going on McCartney so we were back in the air shortly after eight the next morning. The fire had grown very little during the night and still wasn't doing that much when we got over it. I couldn't see anyone on the ground so I tried to reach them on the radio. No one answered. Dillon Dispatch eventually came on to report that no one was probably there to answer yet.

It took a while for that information to soak in. Based on my on the ground experience, over 50 smoke chasers should have arrived on a two acre fire late in the afternoon (as I had seen them do), then taken advantage of the higher humidity, cooler temperatures, and related low fire activity to build a line around the fire that night. They should have eaten the rations and drunk the water they carried and bedded down on the fire line when the fire was adequately controlled for them to do so. Other firefighters could have been called to relieve them at daylight if necessary.

In this case the smoke chasers had fought the fire for an hour or so until about dark then gathered up their equipment and marched off the fire. They ate dinner in a café in Dillon and spent the night camping out in the local KOA Campground. They were still eating breakfast in a local café when I tried to reach them. It was a difficult approach for an old smoke chaser to digest.

I was viewing exactly how personnel management direction has affected the firefighting effort over time. A twelve hour limit on how long a firefighter is supposed to be "on shift" without justification is one of them. Few personnel specialists have ever seen a wildfire. The twelve hour standard has been on the books for years but we've simply ignored it during initial attack if there is any chance of catching the fire. Initial attack firefighters normally start their day doing routine work, such as repairing equipment or driving around looking for fires. They are still reasonably rested when sent on a fire. The adrenalin rush that goes with initial attack is a strong incentive to do what needs to be done, and past training and experience should allow them to do so safely. Seeking forgiveness always seemed easier than losing a fire because of a book standard established by people who don't know what's going on. I always figured it was safer to put in a 24 hour shift to control a small fire before it develops into a major conflagration then it is to get lots of rest and have to face an inferno in the morning. It is easy to blame accidents and fatalities on fatigue rather than poor judgment. Some people will think I am wrong.

The McCartney Fire eventually burned over 30 acres and over 100 firefighters were involved in fighting it. Firefighters and equipment were tied up there when it was needed elsewhere.

The flights got increasingly rough as the season wore on with several flights where we simply could not get into some areas because of smoke, high winds and turbulence. Sometimes it felt like the down drafts were going to rip the wings off the plane. The "stall alarm" kept squawking every time we hit an up or down draft and we spent a lot of time listening to it.

We tried to make the best of things and find areas where the winds would let us stay up but it wasn't easy. And all of the time we knew that we had to return to Dillon and face whatever wind and other weather factors that waited when we had to land.

We were essentially blown out of the sky on one rough trip. We bobbed our way back to Dillon only to be advised that we were facing crosswinds on the landing strip that were gusting to 50 miles per hour. The plane was "crabbing" severely as we approached the strip. The wind was blowing so hard that the plane was lined up somewhere between the prevailing wind and the landing strip as we approached.

We were almost on the ground when a particularly severe gust boosted us off of the runway. Dennis hit the throttle and we bounced upward on the gust. Dennis brought us around for another try. His wife was on the radio in their office on the field, doing her best to help us find the ground between gusts. Unfortunately it's impossible to predict wind gusts. He lined up and we made another pass with the same results. Dennis decided that we would try one more pass. If that didn't work we'd see if we had enough fuel to make it to a landing field somewhere that either had less wind or a runway that somehow lined up better with the wind. I was more than ready to find some ground somewhere. My bladder was indicating that if we didn't do so quickly I was going to embarrass myself. Flying for long periods can cause enough pressure and the current situation wasn't helping much. We came in fast. There were gusts but we simply powered through them. We hit the ground hard and were in the middle of about the third major bounce when Dillon Dispatch called on the radio to see where we were. I don't remember what I told them, but they signed off until we completed several more bounces and finally came to a stop. Mother earth never looked better.

I was working around several other things while all of this was going on. Our son, Jay, had just moved to the Assistant Ranger position on the Teton Basin Ranger District out of Driggs, Idaho. He just arrived when his boss took off on a planning assignment on a Type I Overhead Team and essentially spent all summer on fire assignments, leaving Jay in charge. Having a pregnant wife and transfers seem to go together. They had transferred to Driggs in June and Susan was due to deliver in early August. That call came in the middle of the night on August 5: "Mom, we have another beautiful baby boy. Can you come?" So I made the appropriate phone calls to get a substitute aerial spotter and we headed south to meet Ryan Patrick Pence, our fifth grandchild.

Naturally, Ryan turned out to be the best looking new little grandson anywhere. New babies don't play much, but I played with Ryan and big brother Thomas for an afternoon. Then I left Lois with a promise to come get her when things settled down and headed back north to resume duties.

Montana was closed to outdoor recreation use shortly

after Lois returned. Fires danger was just too high. Any carelessly tossed cigarette or spark from an engine's exhaust, vegetation contacting a catalytic converter under a vehicle, or spark from a horse's shod hoof could cause a fire.

The smoke was so thick around Dillon the day the closure was announced that we were unable to fly. So Lois called a group of friends who would have normally been camping, fishing, riding horses or whatever and we had a "Fire Closure" potluck picnic at our house.

And so went the summer. Fall storms finally broke the cycle. It was nice not to have to fly or teach every day. Standards require that a firefighter must either have a fire assignment or additional training to remain qualified for his "red card" positions. I haven't been on a fire assignment or attended training since 1998, the last year I was called upon to fight wildfire on the ground. But maybe that's okay after 44 years of chasing smoke.

Epilogue

Wildfire problems will continue each year in areas experiencing drought conditions with increased intensity and acres and structures will burn as fuels continue to accumulate under the current political situation. There are no easy answers. I could expend significant pages expounding on the need for sound wild land management by trained professionals instead of management by emotion, political action and court mandate, but I won't.

Hopefully, the agencies can initiate a more aggressive controlled burning program and other management practices will reduce the fire hazard and save money in the long run. Management needs to include measures such as logging and thinning of timber stands to reduce fire intensity where feasible. Congress can correct past mistakes and allow professional land management to return. Don't expect political or judicial miracles.

During very dry summers, we'll run out of qualified

smoke chasers long before we run out of fire. The middle-of-the-night calls will continue on phones everywhere. The Fellowship of Fire will continue as young men and women answer the call. Fire fighters will get so involved in controlling fires that they will overlook known dangers and some of them will die.

And so the Fellowship of Fire continues without me. Still, there's something about the smell of smoke from a wildfire and the Fellowship that keeps calling. Others who have been there know what I mean.

Appendix I: Fire Safety

Rather than tie up a significant part of a chapter with details on the Ten Standard Fire Fighting Orders and Situations that Shout Watch Out, they have been relegated to this appendix. In a way, they would have interrupted the story up front.

The Ten Standard Fire Fighting Orders and the original ten Situations that Shout Watch Out resulted from investigations into what caused the death of twelve smoke jumpers and a smoke chaser on the Mann Gulch Fire on Montana's Helena National Forest 1949.

They have been reworded somewhat from how they were stated when I first memorized them in 1957. There were only ten "Situations that Shout Watch Out" in 1957. There are eighteen "Watch Out Situations" listed in the current Fireline Handbook.

A segment titled "Common Denominators of Fire Behavior on Tragedy Fires" has been added. The additional eight Watch Out Situations and the Common Denominators have been added following investigations into fireline fatalities since Mann Gulch. This information is currently listed just inside the front and back covers of the National Fireline Handbook that is issued to every qualified wildland firefighter and is emphasized over and over as part of their training. Every firefighter is trained to know that his safety, and the safety of his fellow firefighters, is his first responsibility.

Conditions may exist that can lead a firefighter or a crew to violate any of them. I have violated most of them at times and could add pages explaining why I did so. Every firefighter is instructed to consider the consequences if anything goes wrong and to have a safe alternative all figured out before they move into the obviously hazardous situation presented.

FIRE ORDERS

1. Fight fire aggressively but provide for safety first.
2. Initiate all action based on current and expected fire behavior.
3. Recognize current weather conditions and obtain forecasts.
4. Ensure instructions are given and understood.
5. Obtain current information on fire status.
6. Remain in communication with crew members, your supervisor, and adjoining forces.
7. Determine safety zones and escape routes.
8. Establish lookouts in potentially hazardous situations.
9. Retain control at all times.
10. Stay alert, keep calm, think clearly, act decisively.

WATCH OUT SITUATIONS

1. Fire not scouted and sized up.
2. In country not seen in daylight
3. Safety zones and escape routes not identified.
4. Unfamiliar with weather and local factors influencing fire behavior.
5. Uninformed on strategy, tactics, and hazards.
6. Instructions and assignments not clear.
7. No communication link with crew members or supervisor.
8. Constructing line without safe anchoring point.
9. Building fireline downhill with fire below.
10. Attempting frontal assault on fire.
11. Unburned fuel between you and fire.
12. Cannot see main fire, not in contact with someone who can.
13. On a hillside where rolling material can ignite fuel below.
14. Weather becoming hotter and drier.
15. Wind increases and/or changes direction.
16. Getting frequent spot fires across line.
17. Terrain and fuels make escape to safety zones difficult.
18. Taking a nap near fireline.

COMMON DENOMINATORS OF FIRE BEHAVIOR ON TRAGEDY FIRES

1. Most incidents happen on the smaller fires or on isolated portions of larger fires.
2. Most fires are innocent in appearance before the "flare-ups" or "blow-ups." In some cases, tragedies occur in the mop-up stage.

3. Flare-ups generally occur in deceptively light fuels.
4. Fires run uphill surprisingly fast in chimneys, gullies, and on steep slopes.
5. Some suppression tools, such as helicopters or air tankers, can adversely affect fire behavior. The blasts of air from low flying helicopters and air tankers have been known to cause flare-ups.

I would add another concern somewhere above.

OVERCONFIDENCE KILLS!

CPSIA information can be obtained at www.ICGtesting.com
263355BV00005B/50/P

9 781432 767938